Computer Programs for Building Cost Appraisal

Computer Programs for Building Cost Appraisal

Peter S. Brandon
R. Geoffrey Moore
and
Phillip R. Main

COLLINS
8 Grafton Street, London W1

Collins Professional and Technical Books
William Collins Sons & Co. Ltd
8 Grafton Street, London W1X 3LA

First published in Great Britain by
Collins Professional and Technical Books 1985

Distributed in the United States of America
by Sheridan House, Inc.

ISBN 0-00-383043-8

Printed and bound in Great Britain by
Mackays of Chatham, Kent

DISCLAIMER
The publishers and the authors disclaim all responsibility for use of the programs included in this book for any purposes whatsoever. They are general rather than specific examples only. It is therefore the responsibility of the reader to check the suitability of the program and the calculations, the technique and the methodology adopted for the job to which he is applying it before implementing the results.

To	Mary	Audrey	Beverley
	Samantha	Andrew	Matthew
	Juliette	Joanna	
	Robin		

With thanks for their patience and understanding.

Contents

Preface

The success of our previous book, *Microcomputers in Building Appraisal* (Brandon and Moore, Granada 1983) has suggested that there is a demand for some straightforward programs in BASIC related to building cost. Much interest has been shown in the 'starter pack' programs included in that volume and it was suggested that programs of this sort could provide a useful contribution to helping practitioners in their offices and on their home computers. In addition it would be useful for students of quantity surveying and building who needed some practical application for their computer studies.

The problem was which techniques to include in a book of this nature and which to develop in detail. It was decided to take the standard text as *Cost Planning of Buildings*, 5th edition (Ferry and Brandon, Granada 1984), and to select from that publication those techniques which were of practical and research interest and which were not already included in the previous, more general book. Most chapters have therefore a cross reference to *Cost Planning of Buildings*, where more detail on the technique can be found. However, sufficient information has been given in this text for at least an introduction to the subject and a revision for those who have tackled the discipline previously.

In addition, one or two other areas such as unit price estimating and simple statistical measures have been included to broaden the scope of the book.

The purpose of the volume is therefore:

1. To develop the reader's knowledge of programming, in advance of *Microcomputers in Building Appraisal*, by means of practical examples.
2. To assist in the understanding of cost techniques through the program structure.
3. To provide an experimental vehicle for trying out cost techniques on a computer.

4. To provide a base from which practical programs capable of being used in the office can be derived. Some are already in a form for use but our experience has shown that most practitioners wish to tailor a particular format, presentation or technique to their own requirements.

As an introduction to the subject we have dealt with programming to a greater depth in the first chapter than in the remainder of the book.

As far as possible the programs have been made universal for all common microcomputers with an explanation of the anomalies being included in Chapter 11. The major differences tend to occur with the use of peripheral devices such as disks and printers. To avoid too many of these problems the use of these devices has been excluded except where absolutely necessary. We recommend that where differences occur the reader should check with the handbook of his own machine.

It is not absolutely necessary for the reader to understand programming to make use of the programs in this book. However, if all the objectives outlined above are to be achieved then the reader will need to be conversant with at least the rudiments of BASIC.

P. S. Brandon
R. G. Moore
P. R. Main

Acknowledgements

The authors wish to acknowledge the kind help and assistance provided by the following persons without whom this book would not have been possible.

Ruth Hawkesworth, for the illustrations
Eileen Dobson, Audrey Morford, Pat Curtis, Janet Wiles and Christine Bliss, for the secretarial support
Colin Chambers and John Smith, for the technical support
Our families, for their continuing support

We also acknowledge the help and assistance given by many individuals in conversation, discussion and lecture, which contributed indirectly but importantly to the content of this book.

Starter Programs

The programs included in this book are available for the CBM 4032, CBM 8032, Commodore 64, Apple II, BBC, IBM Personal Computer, Intertec Superbrain and Wang Personal Computer machines on disk only.

The price of each disk on which all the programs are recorded is £75 including postage and packing. Please order from:

COED Systems
c/o Collins Professional and Technical Books
8 Grafton Street
London
W1X 3LA
Great Britain

Please enclose:

1. A cheque or money order for the full value, made out to COED Systems.
2. Precise details of your machine, including manufacturer's name and the number and type of disk including the number of tracks.

Chapter One
Simple Statistics

Anyone with an interest in cost advice will be aware that the subject can involve the use of a number of mathematical formulae - for example, statistical calculations, discounting and interest, regression analysis etc. Some of these calculations are quite simple, others more complex. In the first chapter we shall deal with the simple statistical calculations which may be required. It is possible to obtain pocket calculators which have built in programs to calculate the mean and standard deviation of a list of data keyed in and for small lists of data this may well be adequate. One of the problems with such a calculator is that, unless the machine is equipped with a printout facility, there is no record of the data keyed in, so that errors could go undetected. The answer is, of course, to do the calculation twice and see if the same answer is obtained each time.

Any pocket calculator can be used to calculate the arithmetic mean of a group of numbers but a calculator having more than the basic four functions of add, subtract, multiply and divide is required for any other statistical calculation and for many of them it will also be necessary to have one or two memories.

The computer is simply a more advanced calculator which has the advantage of a screen display and therefore can show a considerable amount of information at one time, instead of the single line display of the calculator. It is quite simple to write a computer program which will do the same as a pocket calculator with the advantages that:

1. Instructions to the operator can be printed on the screen and
2. All, or at least many of the numbers typed in can be made to remain on the screen so that numbers which have been input can be checked.

Program structure

A computer program will normally consist of three parts.

1. Inputting the data.
2. Performing the calculations.
3. Printing the results.

A simple program in BASIC which illustrates this can be used to calculate an arithmetic mean of a series of numbers and could be:

```
10 PRINT "HOW MANY OBSERVATIONS";
20 INPUT N
30 PRINT "KEY IN NUMBERS"
40 FOR I=1 TO N
50 INPUT Z
60 T=T+Z
70 NEXT I
80 M=T/N
90 PRINT "ARITHMETIC MEAN =";M
100 END
```

If you run this program the words 'HOW MANY OBSERVATIONS' will appear on the screen followed by a '?' produced by the INPUT statement on line 20. Omitting the semicolon at the end of line 10 would cause the question mark to appear on the next line, i.e. under the 'H' of HOW. On some computers it is possible to include a text string in an INPUT statement in which case lines 10 and 20 would be combined into the single statement:

```
20 INPUT "HOW MANY OBSERVATIONS";N
```

When a number is entered and the RETURN key is pressed, the words 'KEY IN NUMBERS' will appear with a question mark on the line below the letter 'K'. The loop which is set up in line 40 allows the INPUT of N numbers (the RETURN key must be touched after each number has been keyed in). Line 60 adds the numbers together as they are entered. Note that no provision has been made in this program to make sure that the total T is zero at the start. The reason for this is that whenever a program is RUN, all the variables are automatically set to zero by the computer. If we add the option to re-run the program without typing in the word 'RUN' then we must set the variable T to zero before it is re-used. To do this we need to add the following lines to our program:

```
15 T=0
95 PRINT "DO YOU WANT A RE-RUN (Y/N)";
96 INPUT Q$
97 IF Q$="Y" THEN 10
```

Notice that the second part of the program - performing the calculations - is contained in line 60 where the total of the numbers is found and line 80 which calculates the arithmetic mean. Part 3 of the program is contained in line 90.

This program will produce the required results but we are simply using the computer as a desk calculator. The only advantage is that we have some prompts on the screen. If a mistake should be made during entry we have no means of correcting the value.

Another method of entering the data is to use the READ statement and type the DATA into the program before it is RUN. In this way the DATA can be checked and corrected if necessary before the program is used. We can modify the previous program to do this in the following way:

```
20 READ N
40 FOR I=1 TO N
50 READ Z
60 T=T+Z
70 NEXT I
80 M=T/N
90 PRINT "ARITHMETIC MEAN IS";M
```

```
100 END
200 DATA 10
210 DATA 6.5,7.6,6.8,6.9,7
220 DATA 7.5,7.7,6.9,6.8,7.3
```

The DATA statement in line 200 contains the value of N and those in lines 210 and 220 contain the ten values which are to be used. It is important that there are sufficient items contained in these two lines, otherwise the program will stop and an error will be signalled as soon as all the DATA have been used up. If you have entered the program correctly, the value of the arithmetic mean will be shown as 7.1. The number of DATA lines used and the amount of DATA contained in each line is not important. All the numbers used in this example could be entered into line 200. The only limit being the maximum permissible program line length for your machine.

The third method of supplying the program with DATA has been used in Program 1 which uses the GET statement. This statement accepts characters one at a time and by careful programming it is possible to reject any values you wish.

Program 1

The following statistical values can be calculated by simple formulae from any list of numbers:

(a) Arithmetic mean.
(Sum of all the members in the sample divided by the number in the sample)

(b) Geometric mean.
(The nth root of the product of all the members in the sample, where n is the number in the sample)

(c) Standard deviation.
(The deviation of each observation from the arithmetic mean is calculated for each item of data, this is squared to remove negative values, the arithmetic mean of these squared deviations is found and the square root obtained)

(d) Coefficient of variation.

(The standard deviation divided by the arithmetic mean and presented as a percentage)

In order to find the value of the median, the list of data must be sorted into ascending or descending order and is therefore beyond the capacity of an ordinary calculator. A computer can be programmed to sort the data and hence the value of the median can also be obtained.

Program 1, which appears on page 12 is designed to calculate the four values listed above and the median. Because it is necessary to sort the data in order to calculate the median, the data is stored in an array variable L(100) which allows for a maximum of one hundred data values to be input. This array is DIMensioned in line 1015. It is quite simple to increase or decrease the number of values if it should be required, by changing the number within the brackets on line 1015 and the limits of the FOR-NEXT loops in lines 1145 and 1195 together with the maximum value of J in line 1205. The size of the memory on your machine will govern the maximum number of elements which can be contained in this array.

The program is 'menu driven' which means that it begins by displaying a number of choices on the screen and then allows the operator to choose one of these options. In this case we have seven alternatives. The first is to input the data and this is the section which takes up most of the program. Not only are lines 1180-1240 used but also the subroutines in lines 1425-1435 and 6000-6120 are involved, the purpose of each subroutine is shown in REM statements in the program.

The remaining six options are; to calculate the values of arithmetic and geometric means, the standard deviation, the coefficient of variation and the median. The last option is to end the program. The calculation of the various values and the printout of the results on the screen is dealt with in lines 1250-1390. If an attempt is made to choose one of the options two to six before any data have been entered, the message 'NO DATA AVAILABLE' will appear on the screen.

You will notice that the GET statement is used several times in the program. Its first use is for the option selection from the menu in line 1095. It is followed by a routine which checks the value which has been entered. When running this program it is not necessary to touch the RETURN key at this point, in fact the RETURN key along with all the keys other than the numbers 1-7 have no effect. The IF statement on line 1095 then checks which key has been touched and if it is outside the stated range or no key has been touched the program loops back to the GET statement at the beginning of the line. This method of selection can only be used where a single digit reply is expected. When the length of an entry is variable, a more complex routine is required such as that used for the entry of the data later in the program.

Lines 1095-1115 check whether the option chosen is a feasible one. It would be no use selecting options two to six before any data have been entered nor would it be desirable to be able to enter new data if data are already present without first giving the operator the chance of changing his mind. In the simple programs given earlier in the text, the numbers are used as they are INPUT or READ, but in program 1 data are stored in the array variable L(100). A numeric variable, D, is used as an indicator of the state of the numeric array L(100). If the array contains no data, D is set to zero and if it contains data, D is set to 1. This enables the program to select the appropriate message at line 1120 or line 1155 or allow the program to proceed.

Data entry

The program is designed to suit a twenty-five line screen. The first three lines of the data entry section, beginning at line 1180, clear the screen (by using the subroutine at line 4000), carry the message 'PLEASE ENTER YOUR LIST OF DATA' and the instruction on how to terminate the entry sequence by entering the letter 'E' as the value. Fifteen items can be entered using lines four to eighteen on the screen. The number of entries is controlled by the statement in line 1215 which checks that the line number being used on the screen does not exceed eighteen.

The entry of data takes place in the subroutine beginning at line 6000. A description of this subroutine is to be found in Chapter 11.

When fifteen values have been entered or the entry terminated by typing in 'E', an opportunity is given for correcting any of the values displayed on the screen. The same entry subroutine is used to make any alterations which may be necessary and is called by the subroutine in lines 1425-1435.

Once all the corrections have been made, if the entries are not yet completed, the screen is prepared for the next batch of fifteen entries.

When all the data have been entered the program returns to the beginning and displays the MENU. Using option one again will display the message 'THE DATA FILE IS ALREADY FULL' and the choice is given to enter new data. If this option is taken, line 1145 clears all the existing data in the file and sets the variable D to zero to indicate that the file is empty.

Arithmetic mean

This is the second option in the menu and the calculation is carried out in lines 1250-1265. The numeric variable A is used to receive the value of the arithmetic mean. A check is first made, to ascertain whether the variable A has a value. If A is not zero this calculation has been done before and need not be done again and line 1255 is not used. If A is zero, the elements of the array L() are added together using the loop in line 1255 and the mean is found in the same line, the result being printed on the screen in line 1265.

The value of the arithmetic mean is required for the calculation of the standard deviation and the coefficient of variation so that if these options are chosen the program uses this section first and is redirected in line 1260 so that the value of the arithmetic mean is not printed on the screen.

Geometric mean

The third option occurs in lines 1275-1285. The procedure

is similar to that of the arithmetic mean, the calculations taking place in line 1280.

Standard deviation

The fourth option occurs in lines 1295-1310. In order to calculate the standard deviation we begin by finding the sum of the squares of all the numbers using the FOR-NEXT loop in line 1300. The remainder of the calculation takes place in the same line where we divide the result by the number of members in the sample, subtract the arithmetic mean and find the square root.

Coefficient of variation

The fifth option occurs in lines 1320-1325. This requires both the arithmetic mean and the standard deviation to be known so that when this option is selected the route is via these previous two calculations. The calculation of the coefficient of variation takes place in line 1320 and is calculated each time, i.e. no test is made to see if the value V has already been found. The reason for this is that the calculation of the coefficient of variation is simple and is done very quickly, whereas the other calculations involve FOR-NEXT loops which take some time to process and the delay is quite appreciable when a large amount of data have been entered.

Median

The sixth option is the median which is the value of the middle item of a sample - if the number of members in the sample is even, there will be two central values in which case the median is the arithmetic mean of these two central values. Before we can find the central value or values, the numbers must be sorted into ascending or descending order. In this program we have sorted into ascending order and this routine occurs in lines 1335-1370. This is followed by the section of program which calculates and displays the value of the median (lines 1380-1390).

The simplest method of sorting data is the 'exchange

sort' and this is quite satisfactory for small quantities of data. Where the number of values is greater than about thirty the method tends to be rather slow. In this instance we have chosen to use a method known as Shellsort, originally devised by D.L. Shell and considerably quicker than an exchange sort.

To illustrate the method we shall describe its use on a group of ten numbers:

8, 6, 1, 10, 2, 5, 7, 9, 3, 4

Begin by comparing the observations in pairs which are INT(N/2) apart, where N is the number of observations. In this case the value of INT(N/2) is five. If the two observations are not in the correct order they are exchanged.

First pair	8 and 5	change to 5 and 8
Second pair	6 and 7	no change required
Third pair	1 and 9	no change required
Fourth pair	10 and 3	change to 3 and 10
Fifth pair	2 and 4	no change required

After this first pass the list of numbers will read:

5, 6, 1, 3, 2, 8, 7, 9, 10, 4

The gap is now halved, again taking the integer value, in this case two. This time the numbers compared are:

Firstly 5 1 2 7 10
Which will be changed to:
1 2 5 7 10

Secondly 6 3 8 9 4
Which is changed to:
3 4 6 8 9

This makes the list of numbers:

1, 3, 2, 4, 5, 6, 7, 8, 10, 9

Repeating the process reduces the gap to one, which will always be the final gap length, compares all the numbers and results in the sorted list:

1, 2, 3, 4, 5, 6, 7, 8, 9, 10

Having completed the sort, the machine selects the median value - the middle item if the number of observations is odd and the mean of the two central values if the number of observations is even.

Development

You may like to extend this program to produce quartiles, deciles and percentiles.

Calculation of the mode has been omitted from the program because we consider that the best way of showing this is by high resolution graphics and as these vary from one machine to another are not included in any of the programs in this book.

If you wish to include this feature, the following routine can be added which introduces a new array variable MO(100,2) which must be added to the DIM statement in line 1015.

MO(X,1) contains the observation

MO(X,2) contains the number of occurrences

```
1499 REM ******CALCULATE MODE
1500 MO(1,1)=L(1):MO(1,2)=1:K=1
1510 FOR I=2 TO N:IF L(I)<>L(I1) THEN 1530
1520 MO(K,2)=MO(K,2)+1:GOTO 1540
1530 K=K+1:MO(K,1)=L(I):MO(K,2)=1
1540 NEXT I
```

If you find that the memory capacity of your machine is exceeded when the DIM line is encountered it will be necessary to reduce the number of elements in the arrays L() and MO().

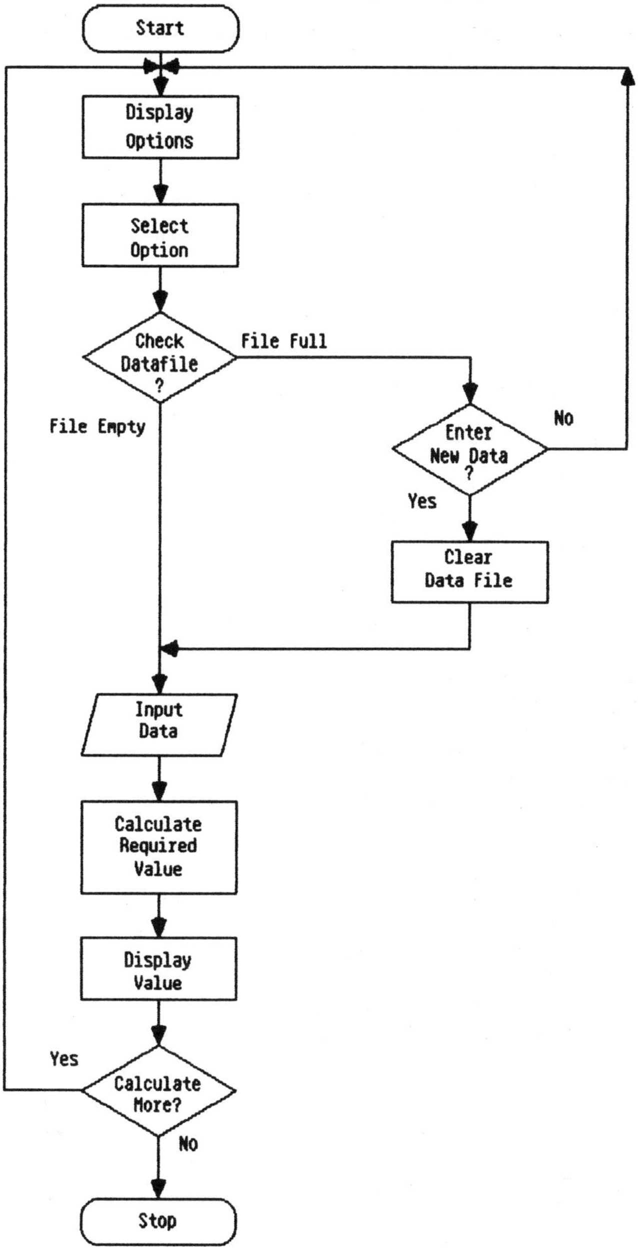

Figure 1.1 Flowchart for Program 1: Simple statistics

```
1000 REM ************** SIMPLE STATISTICS.PROGRAM 1. **************


1005 REM ************  COPYRIGHT BRANDON, MOORE & MAIN ************
1010 REM *COLLINS PROFESSIONAL & TECHNICAL BOOKS (C)84 REV-18/3/84*
1015 DIM L(100):REM                               <DIMENSION ARRAY
1020 SP$="":FOR X= 1 TO 39:SP$=SP$+" ":NEXT X
1025 CS$=CHR$(147):HS$=CHR$(19):RM$="E":DEL$=CHR$(20):CE$="@"
1030 D=0:A=0:G=0:S=0:SO=0


1035 REM **************** DISPLAY MENU OF OPTIONS. ****************
1040 GOSUB 4000:REM                                   CLEAR SCREEN
1045 PRINT TAB(8);"STATISTICAL CALCULATIONS":PRINT
1050 PRINT"THIS PROGRAM HAS THE FOLLOWING OPTIONS":PRINT
1055 PRINT"1) ENTER DATA."
1060 PRINT"2) ARITHMETIC MEAN."
1065 PRINT"3) GEOMETRIC MEAN."
1070 PRINT"4) STANDARD DEVIATION."
1075 PRINT"5) COEFFICIENT OF VARIATION."
1080 PRINT"6) MEDIAN."
1085 PRINT"7) END PROGRAM.":PRINT:PRINT
1090 PRINT"WHICH OPTION DO YOU REQUIRE (1-7)?":PRINT
1095 GOSUB 5500:OP=VAL(Z$):IF OP<1 OR OP>7 THEN 1095
1100 IF OP=7 THEN 9998
1105 IF OP=1 AND D=0 THEN 1180
1110 IF OP>1 AND D=1 THEN 1150
1115 IF OP>1 AND D=0 THEN 1155
1120 PRINT"THE DATA FILE IS ALREADY FULL"
1125 PRINT"DO YOU WISH TO ENTER NEW DATA(Y/N)?"
1130 GOSUB 5500:IF Z$<>"Y" AND Z$<>"N" THEN 1130
1135 IF Z$="N" THEN 1040


1140 REM ******************* CLEAR DATA FILE **********************
1145 FOR J= 1 TO 100:L(J)=0:NEXT J:D=0:A=0:G=0:S=0:SO=0
1150 ON OP GOTO 1180,1250,1275,1250,1250,1335
1155 PRINT:PRINT"NO DATA AVAILABLE."
1160 PRINT:PRINT"PRESS 'RETURN' TO CONTINUE"
1165 GOSUB 5500:IF Z$<>CHR$(13) THEN 1165
1170 GOTO 1040


1175 REM *********************** ENTER DATA ***********************
1180 GOSUB 4000:REM                                   CLEAR SCREEN
1185 PRINT"PLEASE ENTER YOUR LIST OF DATA"
1190 PRINT"TO FINISH ENTER 'E'"
1195 J1=1:N=0:RN=4:FOR J= 1 TO 100
1200 CL=0:GOSUB 5000:PRINT STR$(J);")":D2=1
1205 CL=5:LL=10:I=1:NO=0:GOSUB 6000:IF Z$="E" THEN J=100:GOTO 1230
1210 N=N+1:L(N)=NX:RN=RN+1
1215 IF RN<19 THEN 1230
1220 GOSUB 1400:J1=J1+15:REM                     AMENDMENT ROUTINE
```

```
1225 RN=4:CL=0:GOSUB 5000:FOR X= 1 TO 20:PRINT SP$:NEXT X
1230 NEXT J:D=1
1235 IF Z$="E" OR N=100 THEN GOSUB 1400:REM        AMENDMENT ROUTINE
1240 GOTO 1040


1245 REM **************** CALCULATE ARITHMETIC MEAN ***************
1250 IF A<>0 THEN 1260
1255 FOR J= 1 TO N:A=A+L(J):NEXT J:A=A/N
1260 IF OP=4 OR OP=5 THEN 1295
1265 PRINT"THE ARITHMETIC MEAN IS";A:GOTO 1160


1270 REM ***************** CALCULATE GEOMETRIC MEAN ***************
1275 IF G<>0 THEN 1285
1280 G=1:FOR J= 1 TO N:G=G*L(J):NEXT J:G=G↑(1/N)
1285 PRINT"THE GEOMETRIC MEAN IS";G:GOTO 1160


1290 REM ************** CALCULATE STANDARD DEVIATION **************
1295 IF S<>0 THEN 1305
1300 FOR J= 1 TO N:S=S+L(J)↑2:NEXT J:S=(S/N-A↑2)↑0.5
1305 IF OP=5 THEN 1320
1310 PRINT"THE STANDARD DEVIATION IS";S:GOTO 1160


1315 REM *********** CALCULATE COEFFICIENT OF VARIATION ***********
1320 V=100*S/A
1325 PRINT"THE COEFFICIENT OF VARIATION IS:-":PRINT V;"%":GOTO 1160


1330 REM *********************** SORT DATA ***********************
1335 IF SO=1 THEN 1380
1340 SO=1:G1=N
1345 G1=INT(G1/2):IF G1=0 THEN 1380
1350 FOR I= 1 TO (N-G1):FOR J= I TO 1 STEP (-G1)
1355 J1=J+G1:IF L(J)>L(J1) THEN 1365
1360 J=0:GOTO 1370
1365 T=L(J):L(J)=L(J1):L(J1)=T
1370 NEXT J:NEXT I:GOTO 1345


1375 REM ****************** CALCULATE MEDIAN ********************
1380 IF N/2<>INT(N/2) THEN M1=L((N+1)/2):GOTO 1390
1385 M1=(L(N/2)+L(N/2+1))/2
1390 PRINT"THE MEDIAN IS";M1:GOTO 1160


1395 REM ******************** CHECK SUBROUTINE ******************
1400 RN=22:CL=0
1405 GOSUB 5000:PRINT"ANY ALTERATIONS?(Y/N)                    "
1410 GOSUB 5500:IF Z$<>"Y" AND Z$<>"N" THEN 1410
1415 IF Z$="N" THEN RETURN
```

```
1420 REM ***************** ALTER DATA SUBROUTINE ******************
1425 GOSUB 5000:PRINT"LABEL OF NUMBER TO BE CHANGED":D2=0
1430 CL=32:LL=2:I=0:NO=0:GOSUB 6000:L=NX:IF L<J1 OR L>N THEN 1430
1435 RN=L-J1+4:CL=5:LL=10:I=1:NO=L(L):GOSUB 6000:L(L)=NX:GOTO 1400
```

The following subroutines are required. (See Chapter 11)

4000 Clear screen routine.
4500 Home screen routine.
5000 Cursor position routine.
5500 Input character routine.
6000 Numeric input routine.

Important!
==========
Line 6035 to be changed as follows:

```
6035 IF Z$=RM$ AND D2=1 AND N>2 THEN 6100
```

9998 End program routine.

Chapter Two

Cost Databases

At the heart of nearly all building evaluation programs of any size is a database, that is an organised pool of shareable data usually consisting of regularly updated files which are related together and which permit the direct retrieval of information for a wide range of purposes. Cost evaluation is no exception to this need and indeed in many cases, particularly where there is a need for statistical analysis of the information, it is an absolute necessity. One of the reasons why statistical analysis has not been prevalent in the UK costing procedures is the time consuming nature of undertaking such a task by manual methods.

The subject of databases, their structure, nature and content can be an enormously complex subject and there is a multitude of books on the market to prove it! For the purposes of this book with its limited objectives we have concentrated on providing a simple but useful tool which will allow the computer to file away copies of cost analyses in the UK Building Cost Information concise format. These analyses can then be accessed for any purpose the user may desire and for which he has written the software. For example the user may wish to find all the analyses for brick built offices, or alternatively bring a particular buildings record to the screen and update it using a cost index. On the other hand he may

---

For further reference on this subject see 'Microcomputers in Building Appraisal', Brandon, P.S. and Moore, R.G., Granada 1983. Chapters 4 and 9.

want the arithmetic mean and standard deviation for the cost per square metre of floor area for all primary schools at a common base date of the first quarter 1985. The list of possibilities is practically endless and it remains for the programmer to exercise his skill and imagination to establish his own manipulative programs to make full use of the database provided.

We have not attempted to write more sophisticated manipulation routines but have instead provided a means of creating the database in the first place and then some utility programs to maintain it. The system is written for disk operation and the concise analyses are stored sequentially. This has certain advantages and disadvantages compared with random access. It probably slows the access time down and searching for a single analysis may take longer. However it is simpler, more straightforward and easier to write for several types of machine and we felt that this was the priority of this book. Figure 2.1 shows the content of the database program in diagrammatic form.

Database file construction

The database, as previously mentioned, is constructed using sequential files. Each stored analysis having a sequential file opened under its CI/Sfb code and number. This file holds the costs, general information and description of the analysis. To allow recall and manipulation each analysis has also an entry in an index file called 'BCISDIR'. The entry consists of the analysis identifier followed by some pertinent information to be used if a sort is required on the database.

In general, for speed and efficiency in this type of simple database the index of stored analyses must hold all the information that is likely to be required for sampling from the database without holding specific information such as costs. (Although the contract sum may be of interest.) In other words the index must provide the information for a 'first-level' sort such that a sample may be defined by excluding those analyses which do not match some general parameters. With this in mind the parameters that have been chosen for the index file in

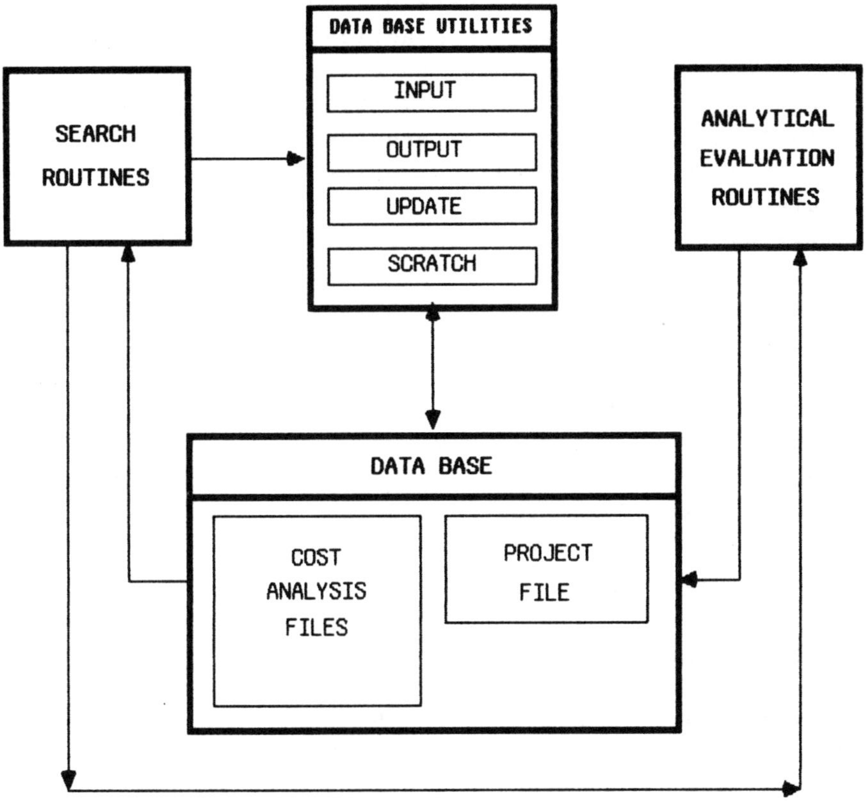

Figure 2.1 Information flow in a typical database suite

this database are: CI/Sfb code, BCIS code (construction type, number of storeys and gross floor area), title, location and date of construction. If these parameters prove insufficient others may be included by altering lines 2235-2240 which take the above information for the analysis and place it in the index array F$(). This must be done before the database is built up otherwise the two index types will be intermixed, rendering the index useless. Some alterations to the directory headings will also be required at lines 1200-1205.

The following is a list of all the parameters held in

an analysis file together with their associated program variables and where appropriate typical examples:

Analysis file breakdown

Indices string:	tender based index'/'cost factor index	IX$
	Example:- 163/192	
Base month:	year '/' month	BM$
	Example:- 78/06	

Element costs

Substructure	B(1)
Superstructure	B(2)
Internal finishes	B(3)
Fittings	B(4)
Services	B(5)
External works	B(6)
Preliminaries	B(7)

Cost information

Measured work	A(1)
PC sums	A(2)
Provisional sums	A(3)
Preliminaries	A(4)
Contingencies	A(5)

Analysis title (max. 21 characters)	JT$
Number of tenders issued '/' received	TE$
Example 05/04	
BCIS code	
Construction type'-'No. of stories'-'Floor area	CO$
Example A-2-2604	

Construction codes

A - Steel framed construction.
B - Brickwork construction.
C - Concrete framed construction.
D - General construction.

Number of stories		NS
Gross floor area		GF
Functional unit		FU
Location	(max. 21 characters)	D$(1)
Client	(max. 21 characters)	D$(2)
Tender dates	(max. 21 characters)	D$(3)
Contract period	(max. 21 characters)	D$(4)
Contract type	(max. 21 characters)	D$(5)
Fluctuations	(max. 21 characters)	D$(6)
Submitted by	(max. 21 characters)	D$(7)

Type of construction

Twelve lines of 35/75 characters describing the project. I$(1-12)

Index file breakdown (Reference: lines 2235-2240)

1. The number of analyses on disk is stored in variable TF.
2. An individual record can be broken down as follows:-

Characters	1 - 7	CI/Sfb code and number.
"	10 - 20	BCIS code.
"	23 - 44	Project title.
"	46 - 67	Location.
"	68 - 73	Base date.

Data storage

In the program no attention has been paid to the coding of the analysis data to maximise disk storage though this might be necessary when using certain disk systems, which have a limited capacity. Certain systems also have constraints as to the maximum number of files permissible on a single disk. Either of these problems may be overcome in one of two ways depending upon the knowledge of the user. The simplest is to use separate disks for major CI/Sfb codes (thus all 'factories' might be placed on one disk) this has the disadvantages of using perhaps many disks and of not allowing global examination across several building types. Another alternative is to code the information ie. by storing large numbers into four characters so that the whole analysis except for the project description could be stored in under 250 bytes (a byte can be thought of as a character for this purpose). The program could then be adapted to use a single random or direct access file instead of the many sequential files and thus make full use of the disk storage available. The programming behind the coding and use of random access files is of a more complex nature, varying considerably between computers and as stated previously, beyond the scope of this book.

Purpose of the program

The program allows the user to:

1. Store, update or erase a concise BCIS analysis on a disk-based computer system, using options 2, 3 and 4 respectively.
2. Display or printout a concise analysis, the numerical data of which can be updated to a user selected tender based and/or cost factor index. (Options 1 and 6.)
3. Display the directory of those analyses with a specified CI/Sfb code using option 5.

Description of Program 2

Lines 1020-1025 Dimension the necessary arrays. These are:

F$() which is the index array (see previous) that is stored in the 'BCISDIR' file. This array is initially set to allow storage of 200 concise analyses. If more than this number can be stored on one disk then it may be suitably increased.

I$() this stores the project description of the analysis being accessed. The length of the strings in this array is dependent upon the number of columns the user selects in line 1045, by presetting variable NL. The maximum length of an element in this array being NL-5. (See line 2220.)

A$(), B$(), C$() are the arrays which hold the standard titles for the general information, cost information and elemental information respectively. These are read in line 1105.

D$() stores the general information of the analysis currently being input/accessed. All elements in this array being subject to a maximum length of twenty-one characters. (See previous)

A() stores the cost information of the analysis currently being input/accessed. (See previous.)

B() stores the elemental cost information of the analysis currently being input/accessed. (See previous.)

Line 1030 Defines a function to take a number (represented by a dummy variable 'X') divide it by the value of the variable GF, which holds the gross floor area, and round off the result to two places of decimals. This function basically finds the cost per metre of an element. The function is called in the program by a statement similar to the one shown below.

Z = FNCPM (Value)

Line 1035 Is the definition of another function to round off a given number to two decimal places.

Line 1040 Is a user defined function which converts a floating point number to an integer by using the INT function. This is then converted to a string variable by the function STR$. Finally the length of this string variable is found by use of the LEN function. Thus the value of 12345.67 would go through the following transformations:

12345.67 - 12345 - " 12345" - 6

Hence the function would return the value of six. This is one more than expected because Commodore BASIC places a space in front of positive numbers. If the user's machine does not use this convention then it may be remedied by adding one to the function thus:-

```
DEF FNB(X) = LEN(STR$(INT(X)))+1
```

Line 1045	Initialises the variable NL to the number of columns displayed on the computer. This figure may be altered to suit the user's machine.
Lines 1050-1060	Initialise the standard variables used in the input routines. (See Chapter 11.)
Lines 1065-1095	Input the index file called 'BCISDIR' from disk.
Line 1105	Reads the standard titles into arrays A$(), B$(), C$() from the data statements at line 10000 onwards.
Lines 1115-1170	Display the main menu of options and wait until a selection is made, by the input of a number between 1 and 7. The program branches to the appropriate line number at line 1170.
Lines 1180-1245	Are the coding to display the directory of stored analyses. The user may choose to display a particular CI/Sfb code or to display the whole directory by pressing the return/enter key when prompted for the CI/Sfb code. At present the

display is halted after every sixth analysis to allow perusal, this number may be increased by altering lines 1230 and 1240 appropriately if the user has an eighty column display.

Lines 1255-1300 Erase an analysis from the disk.

At line 1265 subroutine 2490 is called to input a CI/Sfb code and number. The index array is then checked for the existence of this analysis. If it is not found then an error message is displayed via the routine at 7500 and the user prompted for another code. Once found, the entry in the index of the analyses is deleted using line 1280 and the sequential file relating to the analysis erased from the disk at lines 1290 to 1295. The subroutine 2405 writes the amended index file back to disk and the user is then prompted to press the space bar by the subroutine at line 9000.

Lines 1310-1680 Print an analysis to the printer.

In line 1265 a subroutine is called to load a specified analysis in from disk. The user is then given a choice as to how much of the information is to be printed. By input of the range 1 to 3 the whole file is printed, any one option may be printed by input of, for instance 'FROM 1 TO 1'.

Option 1 prints the general information on the analysis.

Option 2 prints the elemental cost breakdown of the analysis.

Option 3 prints the project description.

Line 1370 opens a printer channel (may or may not be necessary depending on the system being used) followed by a summary of the analysis in lines 1375-1430. A branching routine at lines 1435-1445 then prints the selected options. When finished the printer channel is closed and the program returned to the main menu at line 1450.

Option 1 coding at lines 1455-1520

Option 2 coding at lines 1525-1655

Option 3 coding at lines 1660-1680

Lines 1690-2015 Display a selected analysis. The sub-routine at line 2290 loads the specified analysis into the memory, from the disk. The coding from line 1705 onwards is also used by the update analysis option to display the information before amendment. This is done by the use of a flag variable ZZ.

If the flag is set to zero then after each screenful of information has been displayed subroutine 9000 is called to pause the display until the 'space bar' is pressed and then continue with the display.

If the flag is set to more than 1, then after each screenful of information the program 'returns' to the appropriate amendment routine.

The coding for each screen is as follows:

Screen 1 lines 1705-1740 General information 1.

Screen 2 lines 1745-1810 General information 2.

Screen 3 lines 1815-1870 Cost in-information.

Screen 4 lines 1875-1940 Elemental cost breakdown

lines 1945-1985 Elemental cost per square metre.

Screen 5 lines 1990-2015 Project description.

Lines 2025-2245 Update a specified analysis.
As with the display option the analysis is selected and loaded via the routine at 2290.
As previously stated this option uses the previous options coding to display the information prior to amendment.
Each screen is amended by the following lines.
Screen 1 lines 2040-2060
Screen 2 lines 2065-2130
Screen 3 lines 2135-2160
Screen 4 lines 2165-2200
Screen 5 lines 2205-2225
After screen 5 has been amended the new information is ready to be saved to the disk. If the information just entered is concerned with a new analysis then the total file count is incremented at line 2230. Lines 2235 to 2240 make up the entry of the analysis in the index array. The index file is then updated and the new analysis file written to disk at line 2245.

Lines 2255-2280 Enter a new analysis to be stored.
This routine first calls the subroutine at line 2490 to input a CI/Sfb code and number. The entered code is then compared with the index array to ensure that an analysis does not already exist with this code. The routine then branches to the update analysis routine located at line 2040 onwards. Users of certain computers, notably the BBC microcomputer, may find that it is necessary to initialise some variables before branching to the

update routine. The following extra line will cope with this problem.

```
2267 IX$="":BM$="":JT$="":TE$=""
     CO$="":NS=0:GF=0:FU=0
```

Subroutines

Lines 2290-2395 Load a specified analysis from the disk. If the analysis is to be amended, i.e. OP=3, then the program jumps to the return from subroutine at line 2395. (See line 2355.) If the selected option is to display or print an analysis then the facility is given to update the analysis costs by user specified Tender Price and Cost Factor indices. The default is to show the analysis with its original indices, this is done by typing 'RETURN' over the indices displayed. To enter new indices, clear the entry (see Chapter 11 - Input routines) and enter the updated indices in the form: Tender Based '/' Cost Factor, an example being: 230/201.
Indices outside the range of 100 to 400 are considered to be an input error, the user being prompted for a corrected input. (See lines 2085 and 2375.) The program then proceeds to update the costs accordingly in lines 2380-2390.

Lines 2405-2425 Write the index file 'BCISDIR' to disk. Note that at line 2420 each element of the index array is sent to disk preceded by a character thirty-four (CHR$(34)), this character represents a quotation mark (") which stops the computer from ignoring leading spaces when the file is re-read.

Lines 2435-2480	Write an analysis to the disk under its CI/Sfb code and number. (See line 2440.) A typical example of a file-name would be: 'BCIS 32/ 1' The file-names can be shortened by replacing the 'BCIS' part of the file name to a single character* thus: 'B 32/ 1' at lines 2310 and 2440 if the user's computer has a restriction to the number of characters in a file-name. * or none
Lines 2490-2505	Are the routine to input a CI/Sfb code and number. These being placed in variable NA$, of seven characters in length. Example (as above): ' 32/ 1'
Lines 10000-	Are the data statements for the standard headings.

Important

When using this program for the first time it may be necessary to set up the index file 'BCISDIR' before running the program otherwise a 'file not found' or similar error may be incurred. To avoid this problem a small program needs to be typed in and executed independently. This opens the file 'BCISDIR' for writing, writes a zero to it and closes the file. Some examples for different machines follow below:

Commodore

```
1 DOPEN#1,"BCISDIR",W:PRINT#1,0;CHR$(13);:DCLOSE#1
```

or for BASIC 2.0 users

```
1 OPEN1,8,5,"0:BCISDIR,S,W":PRINT#1,0;CHR$(13);:CLOSE1
```

Apple

```
1 PRINT CHR$(4);"OPEN BCISDIR"
2 PRINT CHR$(4);"WRITE BCISDIR"
3 PRINT 0;CHR$(13);
4 PRINT CHR$(4);"CLOSE BCISDIR"
```

BBC

```
1 X=OPENOUT("BCISDIR")
2 PRINT#X,0
3 CLOSE#X
```

Microsoft BASIC

```
1 OPEN"O",1,"A:BCISDIR"
2 PRINT#1,0;CHR$(13);
3 CLOSE#1
```

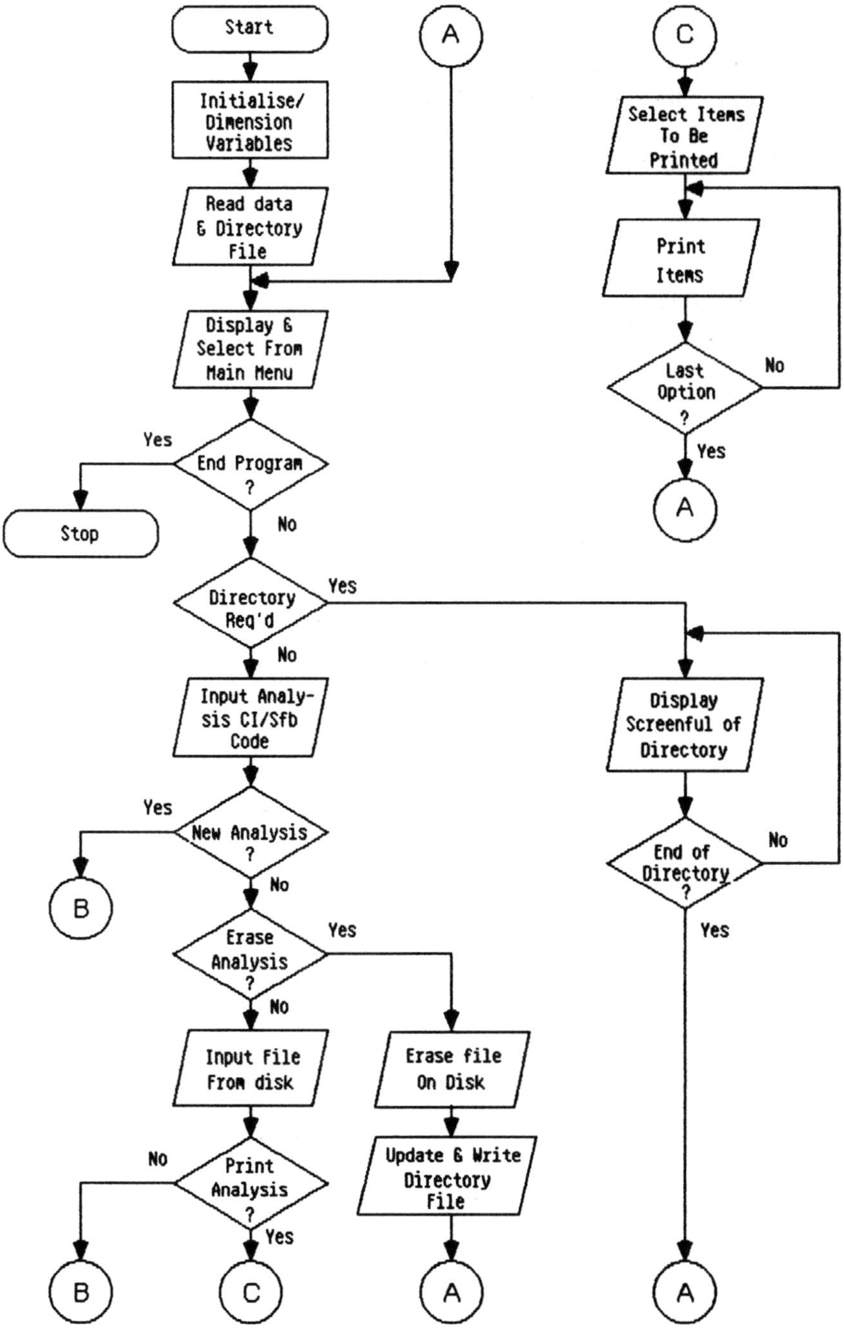

Figure 2.2 Flowchart for Program 2: BCIS analysis database

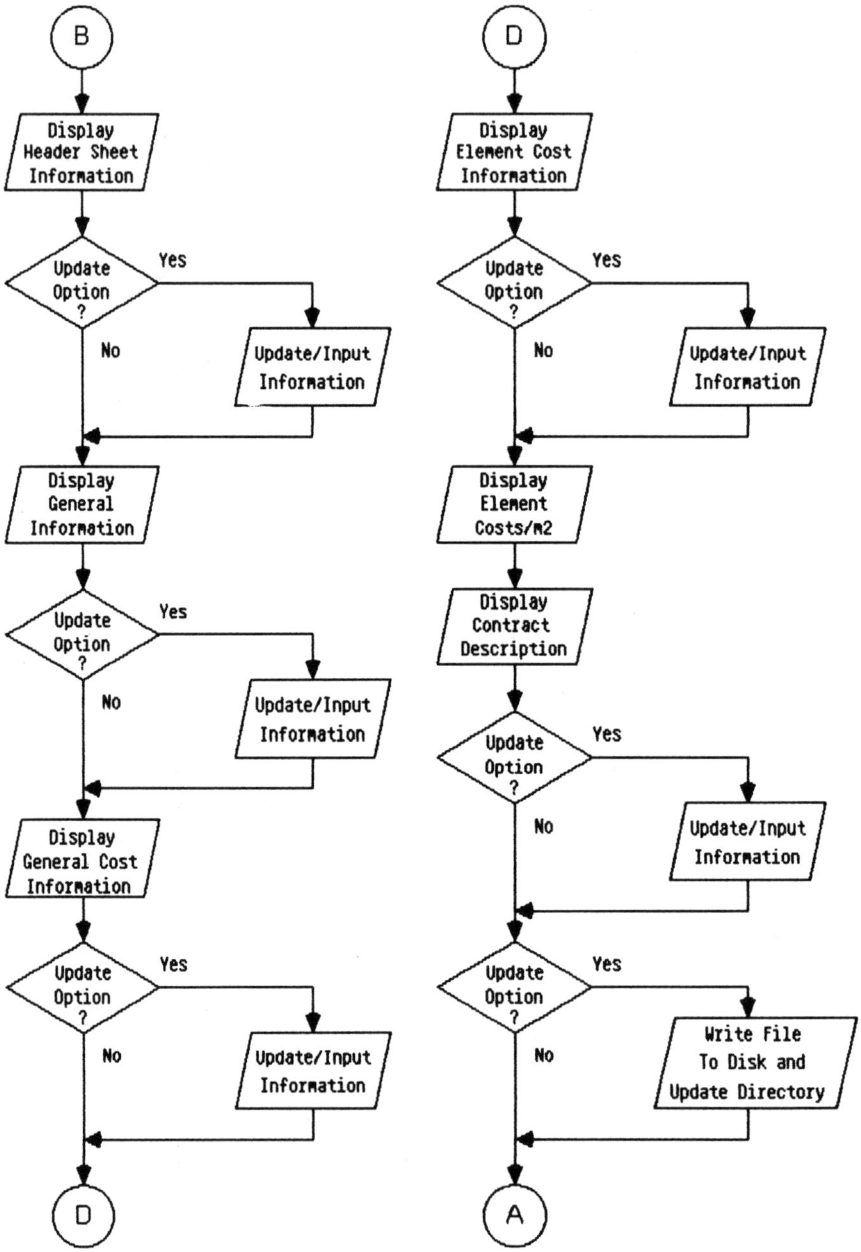

Figure 2.2 contd.

```
1000 REM ********* BCIS ANALYSIS DATABASE. PROGRAM 2. ************

1005 REM *********** COPYRIGHT BRANDON, MOORE & MAIN *************
1010 REM *COLLINS PROFESSIONAL & TECHNICAL BOOKS (C)84 REV12/2/85*

1015 REM ********** INITIALISE VARIABLES/DEFINE ARRAYS ***********
1020 DIM F$(200),I$(12),A$(7),B$(7),C$(7),D$(7),A(7),B(7),C(7)
1025 DIM N(6),N$(3)
1030 DEF FNCPM(X)=INT(0.5+100*(X/GF))/100:REM CONVERT TO COST/M2
1035 DEF FNA(X)=INT(.5+100*X)/100:REM          ROUND OFF NO.TO 2 DPS
1040 DEF FNB(X)=LEN(STR$(INT(X))):REM          FIND LENGTH OF NUMBER
1045 NL=40:REM**** NO. OF COLUMNS ON CO
1050 LT$="":SP$="":FOR X= 2 TO NL:LT$=LT$+"-":SP$=SP$+" ":NEXT X
1055 CS$=CHR$(147):HS$=CHR$(19):RM$="↑":FC$="!":RV$="<":RO$=">"
1060 DEL$=CHR$(20):CE$="@":LF$="":RT$=CHR$(13)
1065 GOSUB 4000:REM                                   CLEARS SCREEN
1070 PRINT"PLEASE WAIT WHILE THE DATA IS READ..."

1075 REM ************* READ DIRECTORY OF ANALYSES ****************
1080 DOPEN#1,"BCISDIR",D0:REM                     OPEN DISK CHANNEL!!!
1085 INPUT#1,TF:IF TF=0 THEN F$(1)=" NO ANALYSES ON DISK":GOTO 1095
1090 FOR X= 1 TO TF:INPUT#1,F$(X):NEXT X
1095 DCLOSE#1:REM                                CLOSE DISK CHANNEL!!!

1100 REM *************** READ DATA (HEADING TITLES) ***************
1105 FOR X= 1 TO 7:READ A$(X),B$(X),C$(X):NEXT X

1110 REM ***************** MAIN MENU OF OPTIONS *******************
1115 GOSUB 4000:REM                                   CLEARS SCREEN
1120 PRINT"       ***";RV$;"BCIS ANALYSIS PROGRAM";RO$;"***":PRINT
1125 PRINT RV$;"OPTIONS:";RO$:PRINT
1130 PRINT RV$;"1";RO$"-READ ANALYSIS.":PRINT
1135 PRINT RV$;"2";RO$"-INPUT ANALYSIS.":PRINT
1140 PRINT RV$;"3";RO$"-UPDATE ANALYSIS.":PRINT
1145 PRINT RV$;"4";RO$"-ERASE ANALYSIS.":PRINT
1150 PRINT RV$;"5";RO$"-DIRECTORY OF STORED ANALYSES.":PRINT
1155 PRINT RV$;"6";RO$"-PRINTOUT ANALYSIS.":PRINT
1160 PRINT RV$;"7";RO$"-EXIT FROM PROGRAM.":PRINT
1165 GOSUB 5500:OP=VAL(Z$):IF OP=0 OR OP>7 THEN 1165
1170 ON OP GOTO 1690,2255,2025,1255,1180,1310,9998

1175 REM ***************** DIRECTORY OF ANALYSES ******************
1180 QU$="CI/SFB CODE":RN=20:CL=0:LL=4:I=0:NO=0:GOSUB 9500
1185 GOSUB 6000:C$=RIGHT$(SP$+STR$(NX),4):GOTO 1210
1190 GOSUB 4000:REM                                   CLEARS SCREEN
1195 PRINT"         ***";RV$;"DIRECTORY OF ANALYSES";RO$;"***":PRINT
1200 PRINT" CI/SFB  BCIS CODE   TITLE";LEFT$(SP$,17);"LOCATION";
```

```
1205 PRINT LEFT$(SP$,14);"DATE":PRINT LEFT$(LT$,NL-1):RETURN
1210 GOSUB 1190:ZZ=0
1215 FOR X= 1 TO TF
1220 IF LEFT$(F$(X),4)<>C$ AND TF>0 AND NX>0 THEN 1235
1225 PRINT F$(X):PRINT:ZZ=ZZ+1
1230 IF ZZ=6 THEN GOSUB 9000:GOSUB 1190:ZZ=0
1235 NEXT X
1240 IF TF=0 OR TF/6<>INT(TF/6) THEN GOSUB 9000
1245 GOTO 1115


1250 REM ************* ERASE AN ANALYSIS OPTION *****************
1255 GOSUB 4000:REM                                 CLEARS SCREEN
1260 PRINT"      ***";RV$;"ERASE ANALYSIS OPTION";RO$;"***"
1265 GOSUB 2490
1270 FOR X= 1 TO TF:IF NA$=LEFT$(F$(X),7) THEN 1280
1275 NEXT X:QU=28:ER$="ANALYSIS DOES NOT EXIST":GOSUB 7505:GOTO 1265
1280 FOR Y= X TO TF:F$(Y)=F$(Y+1):NEXT Y
1285 PRINT:PRINT:PRINT:PRINT LEFT$(SP$,12);"ANALYSIS BEING ERASED."
1290 DF$="BCIS"+NA$:F$(TF)="":TF=TF-1
1295 SCRATCH(DF$),D0:REM                            SCRATCH FILE!!!
1300 GOSUB 2405:RN=15:GOSUB 5000:GOSUB 9000:Z$=RM$:GOTO 9020


1305 REM *************** PRINT AN ANALYSIS OPTION ****************
1310 GOSUB 4000:REM                                 CLEARS SCREEN
1315 PRINT"     ***";RV$;"PRINT OUT ANALYSIS OPTION";RO$;"***"
1320 GOSUB 2290:PRINT:PRINT
1325 PRINT RV$;"OPTIONS:";RO$:PRINT
1330 PRINT RV$;"1";RO$;"-TOTAL PROJECT DETAILS.":PRINT
1335 PRINT RV$;"2";RO$;"-ANALYSIS OF SINGLE BUILDING.":PRINT
1340 PRINT RV$;"3";RO$;"-GENERAL INFORMATION":PRINT:PRINT"FROM ";
1345 GOSUB 5500:A=VAL(Z$):IF A=0 OR A>3 THEN 1345
1350 PRINT Z$;" TO ";
1355 GOSUB 5500:B=VAL(Z$):IF B=0 OR B<A OR B>3 THEN 1355
1360 PRINT Z$:PRINT
1365 PRINT LEFT$(SP$,10);RV$;"PRINTING IN PROGRESS";RO$
1370 OPEN2,4:L$=FC$:Z$=LF$:REM                 OPEN PRINTER CHANNEL!!!
1375 PRINT#2,LEFT$(SP$,30)"BCIS ANALYSIS PRINTOUT";Z$
1380 PRINT#2,LEFT$(SP$,30);LEFT$(LT$,22);Z$
1385 PRINT#2,"      CI/SFB CODE: ";LEFT$(NA$+SP$,27);
1390 PRINT#2,"TITLE: ";JT$;Z$
1395 PRINT#2,"        BCIS CODE: ";LEFT$(CO$+SP$,18);
1400 PRINT#2,"ACTUAL INDICES: ";IX$;" UPDATED TO: ";IY$;Z$
1405 PRINT#2,"       BASE MONTH: ";LEFT$(BM$+SP$,18);
1410 PRINT#2,"NO. OF STOREYS:";NS;Z$
1415 PRINT#2,"GROSS FLOOR AREA:";GF;LEFT$(SP$,16-LEN(STR$(GF)));
1420 PRINT#2," FUNCTIONAL UNIT:";FU;Z$
1425 PRINT#2,"     SUBMITTED BY: ";D$(7);Z$:QQ=A
1430 PRINT#2,LEFT$(LT$+LT$,79);Z$
1435 ON QQ GOTO 1455,1525,1660
1440 QQ=QQ+1:IF QQ<=B THEN 1430
1445 PRINT#2,LEFT$(LT$+LT$,79);Z$
1450 CLOSE2:Z$=RM$:GOTO 9020:REM               CLOSE PRINTER CHANNEL!!!
1455 PRINT#2,LEFT$(SP$,30);"TOTAL PROJECT DETAILS";Z$
```

```
1460 PRINT#2,LEFT$(LT$+LT$,79);Z$
1465 FO$=CHR$(143)+CHR$(150)+CHR$(143)+CHR$(75)
1470 FOR X= 1 TO 4
1475 N$(1)=A$(X):N$(2)=D$(X):N$(3)=B$(X):N(4)=A(X):GOSUB 8000
1480 NEXT X
1485 TT=A(1)+A(2)+A(3)+A(4)
1490 N$(1)=A$(5):N$(2)=D$(5):N$(3)="SUB-TOTAL:":N(4)=TT:GOSUB 8000
1495 N$(1)=A$(6):N$(2)=D$(6):N$(3)=B$(5):N(4)=A(5):GOSUB 8000
1500 PRINT#2,L$;LEFT$(LT$,16);L$;LEFT$(LT$,23);L$;LEFT$(LT$,16);
1505 PRINT#2,L$;LEFT$(LT$,11);L$;Z$:TT=TT+A(5)
1510 PRINT#2,"TENDERS(ISSUED/RECEIVED): ";LEFT$(TE$+SP$,16);
1515 PRINT#2,"CONTRACT SUM:";RIGHT$(SP$+STR$(TT),14);Z$
1520 GOTO 1440
1525 PRINT#2,LEFT$(SP$,25);"ANALYSIS OF A SINGLE BUILDING";Z$
1530 TT=0:FOR X= 1 TO 6:TT=TT+B(X):NEXT X
1535 FOR X= 1 TO 6:C=B(X)+B(7)*(B(X)/TT):C(X)=FNA(C):NEXT X
1540 X$=LEFT$(LT$,11)
1545 W$=L$+LEFT$(LT$,16)+L$+X$+L$+X$+L$+X$+L$+X$+L$+X$+L$
1550 PRINT#2,W$;Z$
1555 PRINT#2,L$;"      ELEMENT     ";L$;"TOTAL COST ";L$;" COST PER  ";
1560 PRINT#2,L$;"TOTAL COST ";L$;"COST/M2     ";L$;" COST/M2    ";L$Z$
1565 PRINT#2,L$;LEFT$(SP$,16);L$;"OF ELEMENT ";L$;" SQ.METRE  ";L$;
1570 PRINT#2,"INC.PRELIMS";L$;"INC.PRELIMS";L$;"BASE DATE77";L$;Z$
1575 FO$=CHR$(143)+CHR$(75)+CHR$(11)+CHR$(75)+CHR$(11)+CHR$(11)
1580 PRINT#2,W$;Z$:TS=0:FOR Z= 1 TO 5
1585 X=B(Z):Y=C(Z):TS=TS+Y:W=(FNCPM(Y)*I1)/I2:N$(1)=C$(Z)
1590 N(2)=X:N(3)=FNCPM(X):N(4)=Y:N(5)=FNCPM(Y):N(6)=W
1595 GOSUB 8000:NEXT Z:PRINT#2,W$;Z$
1600 TT=TT-B(6):W=(FNCPM(TS)*I1)/I2:N$(1)="SUB-TOTAL="
1605 N(2)=TT:N(3)=FNCPM(TT):N(4)=TS:N(5)=FNCPM(TS):N(6)=W
1610 GOSUB 8000:PRINT#2,W$;Z$
1615 X=B(6):Y=C(6):W=(FNCPM(Y)*I1)/I2:N$(1)=C$(6)
1620 N(2)=X:N(3)=FNCPM(X):N(4)=Y:N(5)=FNCPM(Y):N(6)=W
1625 GOSUB 8000
1630 N$(1)=C$(7):X=B(7):N(2)=X:N(3)=FNCPM(X):N(4)=0:N(5)=0:N(6)=0
1635 GOSUB 8000:PRINT#2,W$;Z$
1640 TT=TT+B(6)+B(7):W=(FNCPM(TT)*I1)/I2:N$(1)="TOTAL EXC.CONT="
1645 N(2)=TT:N(3)=FNCPM(TT):N(4)=TT:N(5)=FNCPM(TT):N(6)=W
1650 GOSUB 8000:PRINT#2,W$;Z$
1655 GOTO 1440
1660 PRINT#2,LEFT$(SP$,30);"TYPE OF CONSTRUCTION";Z$
1665 PRINT#2,LEFT$(LT$+LT$,79);Z$:X=1
1670 PRINT#2,I$(X);:IF NL<=40 THEN X=X+1:PRINT#2," ";I$(X);Z$
1675 PRINT#2,Z$:IF X<12 THEN X=X+1:GOTO 1670
1680 GOTO 1440


1685 REM ***************** READ ANALYSIS OPTION *****************
1690 GOSUB 4000:REM                                  CLEARS SCREEN
1695 PRINT"          ***";RV$;"READ ANALYSIS OPTION";RO$;"***"
1700 GOSUB 2290:ZZ=0
1705 L1$="!"+LEFT$(LT$,15)+"!"+LEFT$(LT$,21)+"!"
1710 RN=6:CL=0:GOSUB 5000:PRINT L1$
1715 PRINT"!JOB TITLE:       !";RIGHT$(SP$+JT$,20);" !"
1720 FOR X= 1 TO 7:PRINT L1$
```

```
1725 PRINT"!";A$(X);"!";RIGHT$(SP$+D$(X),20);" !"
1730 NEXT X:PRINT L1$
1735 IF ZZ>1 THEN RETURN
1740 GOSUB 9000
1745 GOSUB 4000:REM                                    CLEARS SCREEN
1750 PRINT RV$;"GENERAL INFORMATION:";RO$
1755 QU$="BCIS CODE                  ":CL=0:RN=3:LL=10:C$=CO$:GOSUB 9500
1760 QU$="INDICES(FORM:TB/CF)        ":CL=0:RN=6:LL=7:C$=IX$:GOSUB 9500
1765 ZZ$=STR$(NS)
1770 QU$="NO. OF STOREYS             ":CL=0:RN=9:LL=2:C$=ZZ$:GOSUB 9500
1775 ZZ$=STR$(GF)
1780 QU$="GROSS FLOOR AREA           ":CL=0:RN=12:LL=7:C$=ZZ$:GOSUB 9500
1785 ZZ$=STR$(FU)
1790 QU$="FUNCTIONAL UNIT            ":CL=0:RN=15:LL=7:C$=ZZ$:GOSUB 9500
1795 QU$="TENDERS(FORM:ISSD/RECD)":CL=0:RN=18:LL=5:C$=TE$:GOSUB 9500
1800 QU$="BASE MONTH (FORM:YR/MN)":CL=0:RN=21:LL=5:C$=BM$:GOSUB 9500
1805 IF ZZ>1 THEN RETURN
1810 GOSUB 9000
1815 GOSUB 4000:REM                                    CLEARS SCREEN
1820 PRINT RV$;"COST INFORMATION:";RO$:PRINT
1825 L1$=LEFT$(L1$,26)+"!":L2$="!"+LEFT$(LT$,15)+"!"+LEFT$(LT$,9)+"!"
1830 TT=0:PRINT L2$:FOR X= 1 TO 4:TT=TT+A(X)
1835 PRINT"!";B$(X);"!";RIGHT$(SP$+STR$(A(X)),8);" !"
1840 IF X<>4 THEN PRINT L1$
1845 NEXT X:PRINT L2$
1850 PRINT" SUB-TOTAL:    ";LEFT$(SP$,11-FNB(TT));TT
1855 QU$=LEFT$(B$(5),14):CL=0:RN=14:LL=8:C$=STR$(A(5)):GOSUB 9500
1860 PRINT" CONTRACT SUM:";LEFT$(SP$,11-FNB(TT+A(5)));TT+A(5)
1865 IF ZZ>1 THEN RETURN
1870 GOSUB 9000
1875 GOSUB 4000:REM                                    CLEARS SCREEN
1880 PRINT RV$;"ELEMENT COSTS";RO$:PRINT
1885 TT=0:PRINT L2$:FOR X= 1 TO 5:TT=TT+B(X)
1890 PRINT"!";C$(X);"!";RIGHT$(SP$+STR$(B(X)),8);" !":PRINT L1$
1895 NEXT X
1900 PRINT" SUB-TOTAL:    ";LEFT$(SP$,11-FNB(TT));TT
1905 PRINT L2$
1910 PRINT"!";LEFT$(C$(6)+SP$,15);"!";RIGHT$(SP$+STR$(B(6)),8);" !"
1915 PRINT L1$
1920 PRINT"!";LEFT$(C$(7)+SP$,15);"!";RIGHT$(SP$+STR$(B(7)),8);" !"
1925 PRINT L1$:TT=TT+B(6)+B(7)
1930 PRINT" TOTAL:         ";LEFT$(SP$,11-FNB(TT));TT
1935 IF ZZ>1 THEN RETURN
1940 GOSUB 9000
1945 GOSUB 4500:REM                                     HOMES SCREEN
1950 PRINT TAB(30);RV$;"COST/M2";RO$:PRINT:PRINT
1955 FOR Y= 1 TO 5:X=B(Y):Z=FNCPM(X)
1960 PRINT TAB(35-FNB(Z));Z:PRINT:NEXT Y
1965 X=FNCPM(TT-B(6)-B(7)):PRINT TAB(35-FNB(X));X:PRINT
1970 X=FNCPM(B(6)):PRINT TAB(35-FNB(X));X:PRINT
1975 X=FNCPM(B(7)):PRINT TAB(35-FNB(X));X:PRINT
1980 X=FNCPM(TT):PRINT TAB(35-FNB(X));X
1985 GOSUB 9000
1990 GOSUB 4000:REM                                    CLEARS SCREEN
1995 PRINT RV$;"TYPE OF CONSTRUCTION";RO$:PRINT
```

```
2000 FOR Y= 1 TO 12
2005 PRINT RIGHT$(SP$+STR$(Y),2);" ";RIGHT$(SP$+I$(Y),NL-5)
2010 NEXT Y:IF ZZ>1 THEN RETURN
2015 GOSUB 9000:Z$=RM$:GOTO 9020


2020 REM *************** UPDATE ANALYSIS OPTION *****************
2025 GOSUB 4000:REM                                  CLEARS SCREEN
2030 PRINT"       ***";RV$;"UPDATE ANALYSIS OPTION";RO$;"***"
2035 GOSUB 2290:ZZ=2
2040 GOSUB 1705:QU$=""
2045 RN=7:CL=17:LL=20:C$=JT$:GOSUB 7000:JT$=C$
2050 FOR Y= 1 TO 7
2055 RN=7+2*Y:CL=17:LL=20:C$=D$(Y):GOSUB 7000:D$(Y)=C$
2060 NEXT Y
2065 GOSUB 1740
2070 RN=3:CL=26:LL=10:C$=CO$:GOSUB 7000:CO$=C$
2075 RN=6:CL=26:LL=7:C$=IX$:GOSUB 7000
2080 C1=VAL(LEFT$(C$,3)):C2=VAL(RIGHT$(C$,3))
2085 IF C1<100 OR C1>400 OR C2<100 OR C2>400 THEN 2095
2090 IX$=C$:GOTO 2100
2095 QU=25:QU$="INDICES(FORM:TB/CF)      !":GOSUB 7500:GOTO 2075
2100 RN=9:CL=26:LL=2:I=0:NO=NS:GOSUB 6000:NS=NX
2105 RN=12:CL=26:LL=7:I=0:NO=GF:GOSUB 6000:GF=NX
2110 IF GF>0 THEN 2120
2115 QU=25:QU$="GROSS FLOOR AREA         !":GOSUB 7500:GOTO 2105
2120 RN=15:CL=26:LL=7:I=1:NO=FU:GOSUB 6000:FU=NX
2125 RN=18:CL=26:LL=5:C$=TE$:GOSUB 7000:TE$=C$
2130 RN=21:CL=26:LL=5:C$=BM$:GOSUB 7000:BM$=C$
2135 GOSUB 1810:TT=0
2140 FOR Y= 1 TO 4
2145 RN=2+Y*2:CL=18:LL=7:I=0:NO=A(Y):GOSUB 6000:A(Y)=NX:TT=TT+NX
2150 NEXT Y:PRINT:PRINT" SUB-TOTAL:";LEFT$(SP$,14-LEN(STR$(TT)));TT
2155 RN=14:CL=18:LL=7:I=0:NO=A(Y):GOSUB 6000:A(5)=NX:TT=TT+A(5)
2160 PRINT:PRINT" ";B$(6);LEFT$(SP$,9-LEN(STR$(TT)));TT
2165 GOSUB 1870:TT=0
2170 FOR Y= 1 TO 5
2175 RN=2+Y*2:CL=18:LL=7:I=0:NO=B(Y):GOSUB 6000:B(Y)=NX:TT=TT+NX
2180 NEXT Y:PRINT:PRINT" SUB-TOTAL:";LEFT$(SP$,14-LEN(STR$(TT)));TT
2185 IF TT=0 THEN 2165
2190 FOR Y= 6 TO 7
2195 RN=4+Y*2:CL=18:LL=7:I=0:NO=B(Y):GOSUB 6000:B(Y)=NX:TT=TT+NX
2200 NEXT Y:PRINT:PRINT" TOTAL:     ";LEFT$(SP$,14-LEN(STR$(TT)));TT
2205 GOSUB 1945
2210 FOR Y= 1 TO 12:RN=Y+2:CL=0
2215 GOSUB 5000:PRINT RIGHT$(SP$+STR$(Y),2)
2220 CL=3:LL=NL-5:C$=I$(Y):GOSUB 7000:I$(Y)=C$
2225 NEXT Y
2230 IF ZZ=3 THEN TF=TF+1:NN=TF
2235 F$(NN)=NA$+"  "+LEFT$(CO$+SP$,12)+LEFT$(JT$+SP$,22)
2240 F$(NN)=F$(NN)+LEFT$(D$(1)+SP$,22)+BM$
2245 GOSUB 9000:GOSUB 2405:GOSUB 2435:Z$=RM$:GOTO 9020
```

```
2250 REM **************** NEW ANALYSIS OPTION *******************
2255 GOSUB 4000:REM                                     CLEARS SCREEN
2260 PRINT"        ***";RV$;"NEW ANALYSIS OPTION";RO$;"***":ZZ=3
2265 GOSUB 2490
2270 FOR X= 1 TO TF:IF NA$<>LEFT$(F$(X),7) THEN 2280
2275 QU=28:ER$="ANALYSIS EXISTS":GOSUB 7505:GOTO 2265
2280 NEXT X:GOTO 2040


2285 REM *************** LOAD ANALYSIS ROUTINE ******************
2290 GOSUB 2490
2295 FOR NN= 1 TO TF:IF NA$=LEFT$(F$(NN),7) THEN 2305
2300 NEXT NN:QU=28:ER$="NON-EXISTENT":GOSUB 7505:GOTO 2290
2305 RN=8:CL=6:GOSUB 5000:PRINT"LOADING ANALYSIS FROM DISK..."
2310 DF$="BCIS"+NA$
2315 DOPEN#1,(DF$),D0:REM                        OPEN DISK CHANNEL!!!
2320 INPUT#1,IX$,BM$
2325 FOR X= 1 TO 7:INPUT#1,B(X):NEXT X
2330 FOR X= 1 TO 5:INPUT#1,A(X):NEXT X
2335 INPUT#1,JT$,TE$,FL$,CO$,NS,GF,FU
2340 FOR X= 1 TO 7:INPUT#1,D$(X):NEXT X
2345 FOR X= 1 TO 12:INPUT#1,I$(X):NEXT X
2350 DCLOSE#1:REM                               CLOSE DISK CHANNEL!!!
2355 GOSUB 5000:PRINT LEFT$(SP$,30):IF OP=3 THEN 2395


2360 REM ************** UPDATE COST DATA ROUTINE ****************
2365 RN=7:CL=0:LL=7:QU$="INDICES (FORM:TB/CF) ":GOSUB 9500:C$=IX$
2370 GOSUB 7000:I1=VAL(LEFT$(C$,3)):I2=VAL(RIGHT$(C$,3))
2375 IF I1<100 OR I1>400 OR I2<100 OR I2>400 THEN 2365
2380 IY$=C$:C1=VAL(LEFT$(IX$,3)):C2=VAL(RIGHT$(IX$,3))
2385 FOR X= 1 TO 5:A(X)=INT(.5+A(X)*(I1/C1)*(I2/C2)):NEXT X
2390 FOR X= 1 TO 7:B(X)=INT(.5+B(X)*(I1/C1)*(I2/C2)):NEXT X
2395 RETURN


2400 REM ************* WRITE DIRECTORY TO DISK RTN ***************
2405 RN=16:CL=4:GOSUB 5000:PRINT" *** UPDATING DIRECTORY... *** "
2410 SCRATCH"BCISDIR",D0:DOPEN#1,"BCISDIR",D0,W
2415 PRINT#1,TF;RT$;:IF TF=0 THEN 2425
2420 FOR X= 1 TO TF:PRINT#1,CHR$(34);F$(X);RT$;:NEXT X
2425 DCLOSE#1:RETURN


2430 REM ************** WRITE ANALYSIS TO DISK RTN ***************
2435 RN=16:CL=4:GOSUB 5000:PRINT"  WRITING ANALYSIS TO DISK...   "
2440 DF$="BCIS"+NA$
2445 SCRATCH(DF$),D0:DOPEN#1,(DF$),D0,W:REM    OPEN DISK CHANNEL!!!
2450 PRINT#1,IX$;RT$;BM$;RT$;
2455 FOR X= 1 TO 7:PRINT#1,B(X);RT$;:NEXT X
2460 FOR X= 1 TO 5:PRINT#1,A(X);RT$;:NEXT X
2465 PRINT#1,JT$;RT$;TE$;RT$;FL$;RT$;CO$;RT$;NS;RT$;GF;RT$;FU;RT$;
2470 FOR X= 1 TO 7:PRINT#1,D$(X);RT$;:NEXT X
2475 FOR X= 1 TO 12:PRINT#1,I$(X);RT$;:NEXT X
2480 DCLOSE#1:RETURN:REM                        CLOSE DISK CHANNEL!!!
```

```
2485 REM ************* INPUT CI/SFB CODE AND NUMBER **************
2490 QU$="CI/SFB CODE":RN=4:CL=0:LL=4:C$="":GOSUB 9500:GOSUB 7000
2495 NA$=RIGHT$(SP$+C$,4)
2500 QU$=QU$+"  "+NA$+"   NUMBER":RN=4:CL=0:LL=2:C$="":GOSUB 9500
2505 GOSUB 7000:NA$=NA$+"/"+RIGHT$(SP$+C$,2):RETURN
```

The following subroutines are required. (See Chapter 11)

```
4000 Clear screen routine.
4500 Home  screen routine.
5000 Cursor position routine.
5500 Input character routine.
6000 Numeric input routine.
7000 Alphanumeric input routine.
7500 Report error routine.
8000 Format and print line routine.
9000 Pause routine.
9500 Draw input box routine.
9998 End program routine.
```

```
10000 DATA "LOCATION:        ","MEASURED WORK: ","SUBSTRUCTURE:  "
10005 DATA "CLIENT:          ","P.C. SUMS:     ","SUPERSTRUCTURE:"
10010 DATA "TENDER DATES:    ","PROVSN. SUMS:  ","INT.FINISHES:  "
10015 DATA "CONTRACT PERIOD","PRELIMINARIES: ","FITTINGS:      "
10020 DATA "CONTRACT TYPE:   ","CONTINGENCIES: ","SERVICES:      "
10025 DATA "FLUCTUATIONS:    ","CONTRACT SUM:  ","EXTERNAL WORKS:"
10030 DATA "SUBMITTED BY:    ","NIL            ","PRELIMINARIES: "
```

Screen display 2.1

```
        ***<READ ANALYSIS OPTION>***

!----------------------------!---!
!CI/SFB CODE     99    NUMBER ! 9 !
!----------------------------!---!

!---------------!---------------------!
!JOB TITLE:     !      SCREEN DISPLAY !
!---------------!---------------------!
!LOCATION:      !              LONDON !
!---------------!---------------------!
!CLIENT:        ! COLLINS P & T BOOKS !
!---------------!---------------------!
!TENDER DATES:  !            12/12/84 !
!---------------!---------------------!
!CONTRACT PERIOD!            78 WEEKS !
!---------------!---------------------!
!CONTRACT TYPE: !       STANDARD FORM !
!---------------!---------------------!
!FLUCTUATIONS:  !                  NO !
!---------------!---------------------!
!SUBMITTED BY:  !            P.R.MAIN !
!---------------!---------------------!

     <PRESS 'SPACE BAR' TO CONTINUE>
```

Screen display 2.2

```
<GENERAL INFORMATION:>
!-------------------------!-----------!
!BCIS CODE                !  A-1-3000 !
!-------------------------!-----------!

!-------------------------!--------!
!INDICES(FORM:TB/CF)      !225/200 !
!-------------------------!--------!

!-------------------------!---!
!NO. OF STOREYS           ! 1 !
!-------------------------!---!

!-------------------------!--------!
!GROSS FLOOR AREA         !   3000 !
!-------------------------!--------!

!-------------------------!--------!
!FUNCTIONAL UNIT          !   2500 !
!-------------------------!--------!

!-------------------------!------!
!TENDERS(FORM:ISSD/RECD)  !01/01 !
!-------------------------!------!

!-------------------------!------!
!BASE MONTH (FORM:YR/MN)  !84/12 !
!-------------------------!------!

     <PRESS 'SPACE BAR' TO CONTINUE>
```

```
<COST INFORMATION:>

!---------------!---------!
!MEASURED WORK: !   40500 !
!---------------!---------!
!P.C. SUMS:     !   21780 !
!---------------!---------!
!PROVSN. SUMS:  !    5000 !
!---------------!---------!
!PRELIMINARIES: !   11000 !
!---------------!---------!
 SUB-TOTAL:         78280
!---------------!---------!
!CONTINGENCIES: !    2300 !
!---------------!---------!
 CONTRACT SUM:      80580

    <PRESS 'SPACE BAR' TO CONTINUE>
```

Screen display 2.3

```
<ELEMENT COSTS>

!---------------!---------!
!SUBSTRUCTURE:  !    7000 !
!---------------!---------!
!SUPERSTRUCTURE:!   32000 !
!---------------!---------!
!INT.FINISHES:  !    7000 !
!---------------!---------!
!FITTINGS:      !     750 !
!---------------!---------!
!SERVICES:      !   16000 !
!---------------!---------!
 SUB-TOTAL:         62750
!---------------!---------!
!EXTERNAL WORKS:!    4500 !
!---------------!---------!
!PRELIMINARIES: !   11000 !
!---------------!---------!
 TOTAL:             78250

    <PRESS 'SPACE BAR' TO CONTINUE>
```

Screen display 2.4

Chapter Three

Construction of a Building Cost Index

At the heart of any cost system, which relies on past data to provide a forecast for the present or future, will be a mechanism to update the historic information to take account of movements in price and cost levels.

The most common method is by means of an index which measures the increase or decrease in cost over time and which can be applied very quickly to the outdated information.

Cost indices usually have a base of 100 identifying the reference point or base year from which all other movements will be judged. Therefore an increase in the value of the index to 137 would show the costs had risen by 37% over the time period recorded:

$$\frac{(137 - 100)}{100} \times 100 = 37\% \text{ increase}$$

Not all past information will relate to the base year and therefore the percentage increase between one index value and another needs to be calculated. For example, if a cost analysis in 1980 has an index value of 240 compared with a base date of 1975 of 100 and we wish to update this to the present which may have an index value of 280 then the calculation will be:

For further reference on this subject see 'Cost Planning of Buildings' (5th edition) Ferry, D.J. and Brandon, P.S., Granada 1984. Chapter 10.

$$\frac{(280 - 240)}{240} \times 100 = 16.67\% \text{ increase}$$

In times of rapid inflation the percentage added to a cost plan based on a past analysis is sometimes greater than any single element's percentage! This is why the measurement of the change in price levels is so important to get right.

Several firms prefer to construct their own indices for the simple reason that they know its content, particularly the weightings given to its various components and are therefore able to assess how a change in one aspect has an impact on the total. For example, the effect of a wage award will only affect the labour element of the total index. By entering the new labour cost a new index can be created which allows the general effect of the wage award to be identified.

It would be true to say, however, that most firms rely on published information and the problem here is that it can often be three months out of date. This unavoidable time lag, due to the process of publication and mailing, is now being overcome by online computer based information services such as that recently developed by the Building Cost Information Service. Even with this online facility some firms would prefer to operate their own internal index which they can tailor to their own particular need.

The problem is which type of index should the firm choose and what source data does it have in its possession to keep the index up to date? The choice of index will depend on two factors:

(i) What movement of cost or price am I measuring?
(ii) Have I the time and data to maintain it?

The two major forms used in the building industry are 'factor cost' and 'tender based' indices.

The tender based index relies on re-pricing the major items of a bill of quantities at standard rates, found in a published schedule and comparing the value of these with the value of the same items in the tender document. Since the tender document is representative of one aspect of the market, what this index is doing is to measure the prevailing market conditions. Anomalies in one bill of quantities are dealt with by taking a substantial number of bills over the time period and averaging them out. It is a very good guide to the effect of the market on the price the client has to pay for his building but it does not reflect necessarily the increase in cost of basic resources to the contractor.

As an exercise in computing the tender based index is beyond the scope of this book. Although it is possible to computerise the process it would require large data storage for the standard schedule of rates and a fairly fast processing capability to access the schedule. It may also require some interaction between the operator and the machine when the choice of major items is being made. This is not always straightforward under the rules of the most popular method and consequently, although the principles are simple, we have not included the program here.

The factor cost index on the other hand is more straightforward. As its name suggests it merely takes the major factors or resources of the building, finds out how the cost of each has changed over the time period and then weights each according to its importance to the value of the building. This is sometimes known as the 'basket of goods' method. A representative sample of the items expected in a building is selected and the movement in cost for the notional building is measured by the weighted average of these items. A very much simplified example might be to find the change in cost of a brick wall.

If the indices for the individual resources required are calculated as below, a full index can be calculated for the whole wall:

	Index	Weighting	Extension
Bricks	140	20	2800
Mortar	150	2	300
Labour	125	6	750
			3850

Divided by 28

Average weighted index = 137.5

Table 3.1 Weightings for BCIS cost indices - series II revision

Work category	*Per cent weighting*			
	Steel framed	*Concrete framed*	*Brickwork construction*	*General construction*
1. Demolitions	—	—	—	—
2. Excavation	1.89	1.98	1.43	1.68
3. Hardcore	1.18	0.70	0.94	0.94
4. General piling	—	0.35	—	0.09
5. Sheet steel piling	—	—	—	—
6. Concrete	8.08	10.03	4.63	6.84
7. Reinforcement	2.15	7.71	1.85	3.39
8. Structural precast units	2.60	3.69	1.22	2.18
9. Non-struct, precast components	2.67	5.80	1.17	2.70
10. Formwork	3.53	6.20	1.13	3.00
11. Brickwork & blockwork	8.22	10.20	15.62	12.42
12. Stone	—	—	—	—
13. Asphalt work	1.01	1.82	0.49	0.95
14. Slate & tile roofing	0.90	0.20	3.15	1.85
15. Asbestos Cmt sheeting	0.18	—	0.05	0.07
16. Plastic Ctd steel roofing	3.65	0.16	—	0.95
17. Aluminium sheet roofing	0.59	0.15	0.01	0.19
18. Felt roofing	0.52	0.88	0.50	0.60
19. Felt roofing on metal	3.03	—	0.59	1.05
20. Carpentry, board, flooring	3.17	2.37	7.45	5.11
21. Hardwood flooring	—	0.05	0.06	0.04
22. Tile & sheet flooring	0.95	1.01	1.57	1.28
23. Jointless flooring	—	—	—	—
24. Softwood joinery	3.22	1.60	8.35	5.38
25. Hardwood joinery	0.37	0.44	1.24	0.82
26. Ironmongery	0.77	0.89	1.89	1.36
27. Steelwork	0.04	0.23	0.24	0.19
28. Steel windows/doors	3.11	0.88	0.25	1.12
29. Aluminium windows/doors	0.63	3.82	1.35	1.79
30. Metalwork	1.09	1.50	1.56	1.43
31. Cast iron pipes	0.24	0.30	0.12	0.20
32. Plastic pipes	0.39	0.26	0.96	0.64
33. Copper pipes	0.24	0.51	0.88	0.63
34. Steel pipes	0.30	0.06	0.06	0.12
35. Boilers, pumps, etc.	0.08	0.07	0.67	0.37
36. Sanitary fittings	0.63	0.43	1.87	1.20
37. Insulation	0.23	0.11	0.80	0.48
38. Plastering	0,83	1.16	2.99	1.99
39. Beds & screeds	1.93	2.57	1.58	1.92
40. Dry partitions & linings	0.73	1.54	2.47	1.80
41. Tiling & terrazzo	1.39	1.91	1.28	1.47
42. Suspended ceilings	1.13	2.43	0.55	1.16
43. Glass	2.34	1.00	1.52	1.60
44. Decorations	2.45	1.29	3.40	2.64
Electrical	8.42	8.10	9.52	8.89
Heating ventilation and air conditioning	11.31	11.31	13.70	12.50
Lift	0.82	3.85	0.45	1.39
Structural steel	12.99	0.44	0.44	3.58

By kind permission of the Building Cost Information Service.

There are various problems associated with the choice of weightings allowing for factors such as changes in productivity over time but in essence this is the methodology employed in a factor cost index except that there would be many more items.

Fortunately the derivation of weightings, which can be a very time consuming task, is not necessary if you are prepared to accept one of the published listings. In addition the chore of calculating the index for each trade, element or work section is overcome if the indices of the NEDO price adjustment formulae are used. These are published in the UK at monthly intervals and therefore are able to provide regular source data for the maintenance of the index.

We are grateful to the BCIS for allowing us to publish their list of weightings for steel-framed, concrete-framed and brickwork construction together with the general construction index. These are shown in Table 3.1. Using this source and assuming that the operator has access to the NEDO indices we have written a simple program to calculate and output the various results. The actual calculation of the index found in lines 1160 and 1170 is a simple matter of multiplying the NEDO work category index by the relevant weighting.

NEDO index x NEDO weighting = Extended value

The sum of the extended values is then divided by the total of the weightings (100) to arrive at the average for the whole building. Most of the program lines are not concerned with actually doing the calculation but with presenting the information on the screen and printing it in an acceptable manner.

<u>Description of Program 3</u>

Line 1020	Dimensions the necessary arrays, these are: WC$() which holds the descriptions of the 48 NEDO index categories. ID() stores the current NEDO indices which are supplied by the user each

	month.
	CT(,) is a two dimensional array which holds the weightings of the NEDO categories for each of the four construction types.
	FL(,) is also a two dimensional array that stores the results of the calculations, ie. the extended values (see line 1160) over the four construction types.
	P$() stores the construction type headings.
	T() stores the average weighted index for the four construction types.
Line 1030	Initialises variable NI to 48 - the number of NEDO categories.
Lines 1035-1040	Initialise the general variables to be used in the program (see Chapter 11).
Lines 1045-1060	Read the data into the appropriate arrays.
Lines 1065-1125	Display the main menu of the program. Option 1 allows the entry of the NEDO indices with the calculation of the average weighted indices. Option 2 enables the user to obtain a printout of the weightings being used.
Lines 1130-1170	Control the input/amendment of the NEDO indices. A check is made at line 1155 to see that the index is over 100 otherwise an error message routine is called and the user prompted for another input. Line 1160 does the calculation multiplying the NEDO index by the four construction type weightings, storing the result in FL() and summing the total extensions in array T(). The average weighted extension for the whole building is then found at line 1170 by dividing through by the

total of the weightings (100).

Lines 1175-1235 — Display the sub-option menu. The options available are:

Options 1-4 — Display the results of the calculations for the selected construction type. Once displayed there is an option to print a hardcopy.

Option 5 — Allows amendment of the NEDO indices. Correct values may be re-input by simply typing 'RETURN' over them.

Option 6 — Prints the results for all four construction types directly to the printer, starting with steel-framed construction.

Option 7 — Returns to the main menu.

Lines 1240-1245 — Are the coding for option 6 above. Variable KK holds the construction type number and the subroutine at 1355 is called repeatedly to print the results for each type.

Lines 1250-1345 — Display the NEDO index, weighting and extension for each of the 48 categories of the specified construction type (selected by value of variable KK). The display is paused after every sixth category to allow perusal. The average weighted index is then displayed together with an option for hardcopy by calling the sub-routine at 1355.

Lines 1350-1495 — This is the subroutine which outputs the results for a particular construction type to the printer. As previously stated the type of construction is selected by the value of variable KK.

Lines 1500-1645 — Prints the percentage weightings of the categories for all four types of construction to the printer. This allows the user to obtain a check as to the weightings being used.

Data structure

The data have been constructed to allow for ease of alteration. Each NEDO category description being followed by four percentage weightings, which are for:

1. Steel-framed construction.
2. Concrete-framed constructon.
3. Brickwork construction.
4. General construction.

Note that if the standard weightings used by the BCIS are not applicable then the sum of the new weightings calculated by the firm must total to one hundred otherwise line 1170 will need to be amended. (See also Chapter 11.)

Postscript on printers

Reference print routines at lines 1350-1645. (See also Chapter 11.)

1. Printers that do not line feed automatically can be made to do so by setting variable LF$ to a line feed character ie. LF$=CHR$(10) at line 1040.
2. In the listing a vertical line is represented as an exclamation mark. If your printer has a vertical line in its character set then this may be used by setting variable L$ in lines 1355 and 1515 to the appropriate character. ie. L$=CHR$(95)

Data requirements

The operator will need to have before him:

(i) A list of the current NEDO indices (series II).

(ii) If the standard weightings used by the BCIS are not applicable then lines 1030, 1050-1060 and 10000 onwards will need to be altered to change the work sections and new weightings calculated by the firm. This may also affect the titles.

Although the NEDO format has been chosen, it would be quite easy to adapt the program to take a list of materials (such as those published by the Department of Trade and Industry) where indices are published to obtain a composite index for materials; add an index for labour (which could be calculated in another part of the program); do the same for plant; and then weight these three major sub-sections together to produce the total index.

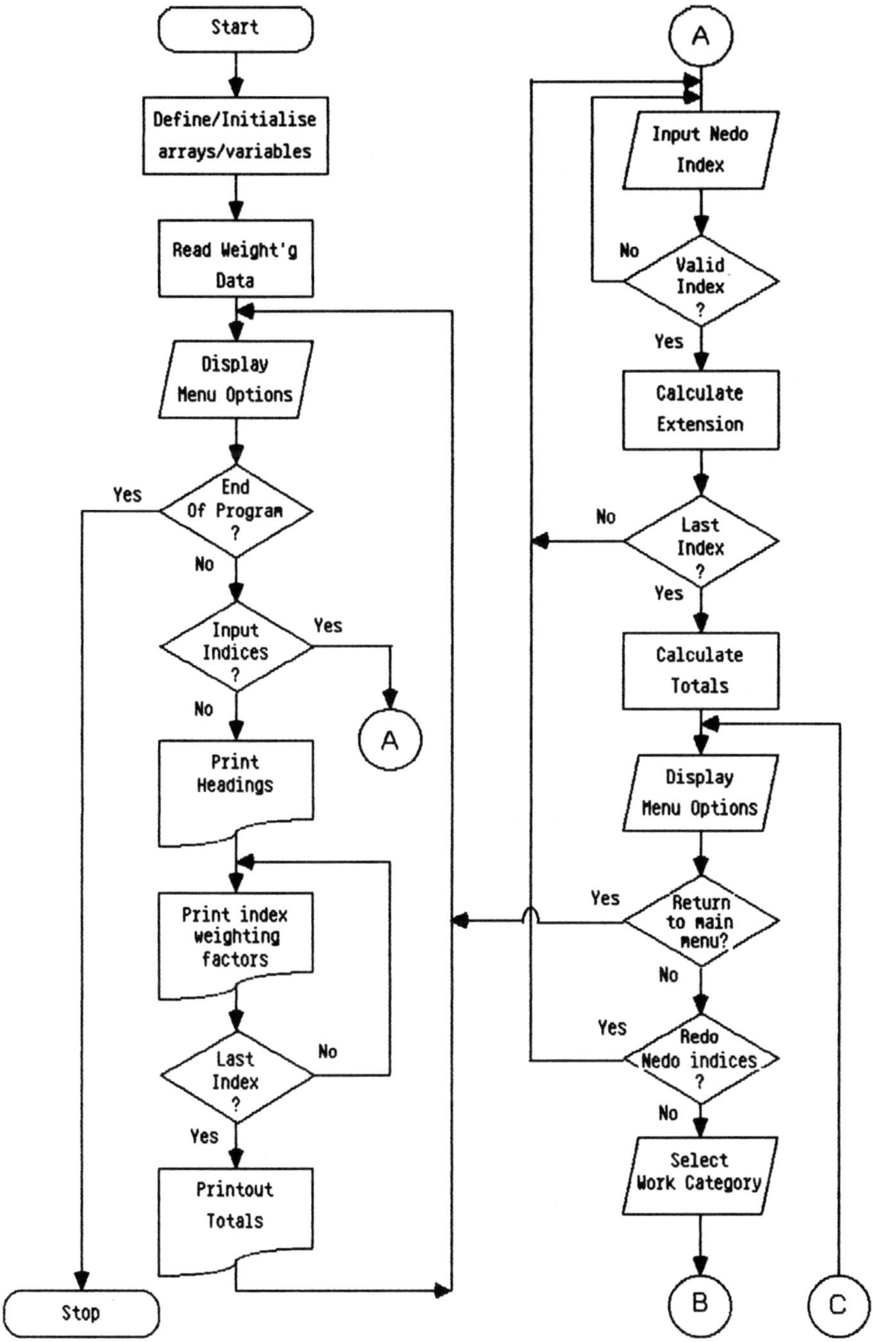

Figure 3.1 Flowchart for Program 3: BCIS indices

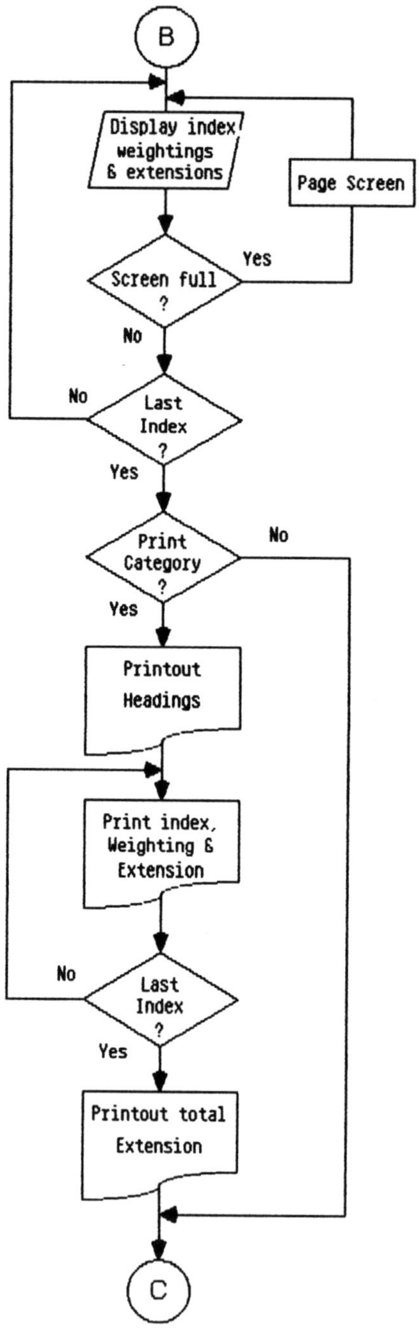

Figure 3.1 contd.

```
1000 REM *************** BCIS INDICES. PROGRAM 3.******************

1005 REM ************ COPYRIGHT BRANDON, MOORE AND MAIN  **********
1010 REM *COLLINS PROFESSIONAL & TECHNICAL BOOKS (C)85 REV 12/2/85*

1015 REM ************* DIMENSION/INITIALISE VARIABLES *************
1020 DIM WC$(50),ID(50),CT(4,50),FL(4,50),P$(4),T(4)
1025 SP$="":LT$="":FOR Z= 1 TO 39:SP$=SP$+" ":LT$=LT$+"-":NEXT Z
1030 NI=48:REM **** NUMBER OF INDICES
1035 CS$=CHR$(147):HS$=CHR$(19):RM$="↑":FC$=" ":RV$="<":RO$=">"
1040 DEL$=CHR$(20):CE$="@":LF$="":PRINT:PRINT"READING THE DATA..."

1045 REM ***************** READ DATA INTO ARRAYS *****************
1050 FOR Z= 1 TO 4:READ P$(Z):NEXT Z
1055 FOR Z= 1 TO NI:READ WC$(Z)
1060 FOR Y= 1 TO 4:READ CT(Y,Z):NEXT Y:NEXT Z

1065 REM ***************** MAIN MENU SELECTION  ******************
1070 GOSUB 4000:REM                                   CLEARS SCREEN
1075 PRINT TAB(3);"*** COST OF PLANNING A BUILDING ***"
1080 PRINT TAB(7);LEFT$(LT$,27)
1085 PRINT TAB(11);"THE B.C.I.S BUILDING"
1090 PRINT TAB(11);LEFT$(LT$,20)
1095 PRINT TAB(15);"COST INDICES":PRINT TAB(15);LEFT$(LT$,12)
1100 PRINT:PRINT"OPTIONS AVAILABLE:-":PRINT
1105 PRINT RV$;"1";RO$;"-ENTER BCIS INDICES."
1110 PRINT RV$;"2";RO$;"-PRINTOUT % WEIGHTINGS."
1115 PRINT RV$;"3";RO$;"-END PROGRAM."
1120 GOSUB 5500:A$=Z$:K=VAL(A$):IF K=0 OR K>3 THEN 1120
1125 ON K GOTO 1135,1505,9998

1130 REM ********** INPUT INDICES & CALCULATE EXTENSIONS **********
1135 RN=14:CL=0:GOSUB 5000:PRINT"INPUT THE NEDO INDEX"
1140 RN=16:FOR T= 1 TO NI:CL=0
1145 GOSUB 5000:PRINT LEFT$(SP$,38):GOSUB 5000:PRINT WC$(T)
1150 LL=3:I=0:CL=30:NO=ID(T):GOSUB 6000
1155 IF NX<100 THEN QU=CL:QU$=WC$(T):GOSUB 7500:GOTO 1150
1160 FOR S= 1 TO 4:FL(S,T)=NX*CT(S,T):T(S)=T(S)+FL(S,T):NEXT S
1165 ID(T)=NX:NEXT T:PRINT:PRINT"INPUT COMPLETE.":GOSUB 9000
1170 FOR S= 1 TO 4:T(S)=T(S)/100:NEXT S

1175 REM ************** SUB-OPTION MENU/SELECTION ****************
1180 GOSUB 4000:REM                                   CLEARS SCREEN
1185 PRINT:PRINT"OPTIONS AVAILABLE:-":PRINT
1190 PRINT RV$;"1";RO$;"-";P$(1)
1195 PRINT RV$;"2";RO$;"-";P$(2)
1200 PRINT RV$;"3";RO$;"-";P$(3)
```

```
1205 PRINT RV$;"4";RO$;"-";P$(4)
1210 PRINT RV$;"5";RO$;"-CHECK BCIS INDICES."
1215 PRINT RV$;"6";RO$;"-PRINTOUT OF ALL OPTIONS."
1220 PRINT RV$;"7";RO$;"-RETURN TO MAIN MENU."
1225 GOSUB 5500:A$=Z$:K=VAL(A$):KK=K:IF K=0 OR K>7 THEN 1225
1230 PRINT:PRINT"OPTION ";A$;" SELECTED."
1235 ON K GOTO 1250,1250,1250,1250,1135,1245,1070


1240 REM ******* PRINTOUT OF ALL CONSTRUCTION TYPE RESULTS *******
1245 FOR KK= 1 TO 4:GOSUB 1355:NEXT KK:GOTO 1180


1250 REM *********** DISPLAY CONSTRUCTION TYPE RESULTS ***********
1255 GOSUB 4000:REM                                        CLEARS SCREEN
1260 PRINT"BUILDING COSTS BASED ON":PRINT P$(K):PRINT
1265 PRINT"     WORK     NEDO       %      EXTENSION"
1270 PRINT"CATEGORY   INDEX   WEIGHTING"
1275 FOR T= 1 TO NI
1280 PRINT:PRINT LEFT$(WC$(T)+SP$,10);RIGHT$(SP$+STR$(ID(T)),6);
1285 B=CT(KK,T):GOSUB 8500:PRINT"  ";N$;
1290 B=FL(KK,T):GOSUB 8500:PRINT"  ";N$
1295 IF T<>NI AND T/6=INT(T/6) THEN GOSUB 9000:RN=6:GOSUB 5000
1300 NEXT T
1305 B=T(KK)*100:GOSUB 8500
1310 PRINT:PRINT"TOTAL EXTENSION IS:-           ";N$
1315 B=(INT(T(KK)*100+0.5))/100:GOSUB 8500:PRINT LEFT$(LT$,28)
1320 PRINT"TOTAL VALUE IS:-    ";N$:PRINT LEFT$(LT$,28)
1325 PRINT"'P' FOR PRINTOUT OR 'C' TO CONTINUE."
1330 GOSUB 5500:A$=Z$:IF A$<>"C" AND A$<>"P" AND A$<>RM$ THEN 1330
1335 IF A$="P" THEN GOSUB 1355
1340 IF A$=RM$ THEN 1070
1345 GOTO 1180


1350 REM ****** PRINTOUT OF RESULTS FOR A CONSTRUCTION TYPE ******
1355 OPEN2,4:Z$=LF$:L$="!":REM               OPEN PRINTER CHANNEL!!!
1360 S$=L$+LEFT$(SP$,23)+L$+LEFT$(SP$,12)+L$+LEFT$(SP$,12)+L$
1365 S$=S$+LEFT$(SP$,14)+L$
1370 H$=L$+LEFT$(LT$,23)+L$+LEFT$(LT$,12)+L$+LEFT$(LT$,12)+L$
1375 H$=H$+LEFT$(LT$,14)+L$
1380 PRINT#2,LEFT$(SP$,20);"BCIS BUILDING COST INDEX";Z$
1385 PRINT#2,LEFT$(SP$,20);LEFT$(LT$,24);Z$:PRINT#2,Z$:PRINT#2,Z$
1390 PRINT#2,LEFT$(SP$,20);"BUILDING COSTS BASED ON:-";Z$
1395 PRINT#2,Z$:PRINT#2,Z$
1400 PRINT#2,LEFT$(SP$,20);P$(KK);Z$
1405 PRINT#2,LEFT$(SP$,20);LEFT$(LT$,LEN(P$(KK)));Z$
1410 PRINT#2,Z$:PRINT#2,Z$:PRINT#2,LEFT$(LT$+LT$,66);Z$
1415 PRINT#2,L$;"      WORK CATEGORY     ";L$;" NEDO INDEX ";L$;
1420 PRINT#2," %WEIGHTING ";L$;"  EXTENSION    ";L$;Z$:PRINT#2,H$;Z$
1425 FOR T= 1 TO NI:PRINT#2,S$;Z$
1430 B=CT(KK,T):GOSUB 8500:M$=N$:B=FL(KK,T):GOSUB 8500
1435 PRINT#2,L$;LEFT$(WC$(T)+SP$,23);L$;RIGHT$(SP$+STR$(ID(T)),8);
1440 PRINT#2,"    ";L$;M$;"    ";L$;N$;"      ";L$;Z$
1445 NEXT T
```

```
1450 PRINT#2,S$;Z$:PRINT#2,H$;Z$:PRINT#2,L$;LEFT$(SP$+SP$,64);L$;Z$
1455 B=T(KK)*100:GOSUB 8500
1460 PRINT#2,L$;"   TOTAL EXTENSION IS:-";LEFT$(SP$,27);N$;
1465 PRINT#2,"      ";L$;Z$:PRINT#2,L$;LEFT$(SP$+SP$,64);L$
1470 PRINT#2,LEFT$(LT$+LT$,66);Z$:PRINT#2,Z$
1475 B=T(KK):GOSUB 8500
1480 PRINT#2,"   TOTAL VALUE IS:-   ";N$;Z$:PRINT#2,Z$
1485 PRINT#2,LEFT$(LT$+LT$,66);Z$:PRINT#2,Z$:PRINT#2,Z$
1490 CLOSE2:REM                              CLOSE PRINTER CHANNEL!!!
1495 RETURN


1500 REM *********** PRINTOUT OF PERCENTAGE WEIGHTINGS ***********
1505 PRINT:PRINT"OPTION 2 SELECTED.":PRINT
1510 PRINT:PRINT"PLEASE WAIT PRINTING IN PROGRESS..."
1515 OPEN2,4:Z$=LF$:L$="!":REM                OPEN PRINTER CHANNEL!!!
1520 PRINT#2,LEFT$(SP$,24);"WEIGHTINGS FOR BCIS COST INDICES";Z$
1525 PRINT#2,LEFT$(SP$,24);LEFT$(LT$,32);Z$
1530 PRINT#2,Z$:PRINT#2,Z$:PRINT#2,LEFT$(LT$+LT$,77);Z$
1535 PRINT#2,L$;LEFT$(SP$,23);L$;LEFT$(SP$+SP$,51);L$;Z$
1540 PRINT#2,L$;LEFT$(SP$,23);L$;"     PER CENT WEIGHTINGS";
1545 PRINT#2,LEFT$(SP$,28);L$;Z$
1550 PRINT#2,L$;"      WORK CATEGORY      ";L$;LEFT$(LT$+LT$,51);L$;Z$
1555 BL$=L$+LEFT$(SP$,23)+L$+LEFT$(SP$,12)+L$+LEFT$(SP$,12)+L$
1560 BL$=BL$+LEFT$(SP$,12)+L$+LEFT$(SP$,12)+L$
1565 HL$=L$+LEFT$(LT$,23)+L$+LEFT$(LT$,12)+L$+LEFT$(LT$,12)+L$
1570 HL$=HL$+LEFT$(LT$,12)+L$+LEFT$(LT$,12)+L$
1575 PRINT#2,BL$;Z$
1580 PRINT#2,L$;LEFT$(SP$,23);L$;"STEEL       ";L$;"CONCRETE    ";
1585 PRINT#2,L$;"BRICKWORK   ";L$;"GENERAL     ";L$;Z$
1590 PRINT#2,L$;LEFT$(SP$,23);L$;"FRAMED      ";L$;"FRAMED      ";
1595 PRINT#2,L$;"CONSTRUCTION";L$;"CONSTRUCTION";L$;Z$
1600 PRINT#2,BL$;Z$:PRINT#2,HL$;Z$
1605 FOR T= 1 TO NI
1610 PRINT#2,BL$;Z$:PRINT#2,L$;LEFT$(WC$(T)+SP$,23);L$;
1615 FOR S= 1 TO 4
1620 B=CT(S,T):GOSUB 8500:PRINT#2,N$;"    ";L$;
1625 NEXT S:PRINT#2,Z$
1630 NEXT T
1635 PRINT#2,HL$;Z$:PRINT#2,Z$:PRINT#2,Z$
1640 CLOSE2:REM                              CLOSE PRINTER CHANNEL!!!
1645 GOTO 1070
```

The following subroutines are required. (See Chapter 11)

4000 Clear screen routine.
4500 Home screen routine.
5000 Cursor position routine.
5500 Input character routine.
6000 Numeric input routine.
7500 Report error routine.
8500 Format number to two decimal places.
9000 Pause routine.
9998 End program routine.

```
10000 REM ******************* DATA STATEMENTS ********************
10005 DATA"STEEL-FRAMED CONSTRUCTION."
10010 DATA"CONCRETE-FRAMED CONSTRUCTION."
10015 DATA"BRICKWORK CONSTRUCTION."
10020 DATA"GENERAL CONSTRUCTION."
10025 DATA"DEMOLITIONS",0,0,0,0
10030 DATA"EXCAVATION",1.89,1.98,1.43,1.68
10035 DATA"HARDCORE",1.18,0.7,0.94,0.94
10040 DATA"GENERAL PILING",0,0.35,0,0.09
10045 DATA"SHEET STEEL PILING",0,0,0,0
10050 DATA"CONCRETE",8.08,10.03,4.63,6.84
10055 DATA"REINFORCEMENT",2.15,7.71,1.85,3.39
10060 DATA"STRUCTURAL PRECAST COMPONENTS",2.6,3.69,1.22,2.18
10065 DATA"NON-STRUCT PRECAST COMPONENTS",2.67,5.8,1.17,2.7
10070 DATA"FORMWORK",3.53,6.2,1.13,3
10075 DATA"BRICKWORK & BLOCKWORK",8.22,10.2,15.62,12.42
10080 DATA"STONE",0,0,0,0
10085 DATA"ASPHALT WORK",1.01,1.82,0.49,0.95
10090 DATA"SLATE & TILE ROOFING",0.9,0.2,3.15,1.85
10095 DATA"ASBESTOS CMT. SHEETING",0.18,0,0.05,0.07
10100 DATA"PLASTIC CTD. STEEL ROOFING",3.65,0.16,0,0.95
10105 DATA"ALUMINIUM SHEET ROOFING",0.59,0.15,0.01,0.19
10110 DATA"FELT ROOFING",0.52,0.88,0.5,0.6
10115 DATA"FELT ROOFING ON METAL",3.03,0,0.59,1.05
10120 DATA"CARPENTRY, BOARD, FLOORING",3.17,2.37,7.45,5.11
10125 DATA"HARDWOOD FLOORING",0,0.05,0.06,0.04
10130 DATA"TILE & SHEET FLOOR",0.95,1.01,1.57,1.28
10135 DATA"JOINTLESS FLOORING",0,0,0,0
10140 DATA"SOFTWOOD JOINERY",3.22,1.6,8.35,5.38
10145 DATA"HARDWOOD JOINERY",0.37,0.44,1.24,0.82
10150 DATA"IRONMONGERY",0.77,0.89,1.89,1.36
10155 DATA"STEELWORK",0.04,0.23,0.24,0.19
10160 DATA"STEEL WINDOW/DOOR",3.11,0.88,0.25,1.12
10165 DATA"AL. WINDOWS/DOORS",0.63,3.82,1.35,1.79
10170 DATA"METALWORK",1.09,1.5,1.56,1.43
10175 DATA"CAST IRON PIPES",0.24,0.3,0.12,0.2
10180 DATA"PLASTIC PIPES",0.39,0.26,0.96,0.64
10185 DATA"COPPER PIPES",0.24,0.51,0.88,0.63
10190 DATA"STEEL PIPES",0.3,0.06,0.06,0.12
10195 DATA"BOILERS, PUMPS ETC.",0.08,0.07,0.67,0.37
10200 DATA"SANITARY FITTINGS",0.63,0.43,1.87,1.2
10205 DATA"INSULATION",0.23,0.11,0.8,0.48
10210 DATA"PLASTERING",0.83,1.16,2.99,1.99
10215 DATA"BEDS & SCREEDS",1.93,2.57,1.58,1.92
10220 DATA"DRY PARTITIONS & LININGS",0.73,1.54,2.47,1.8
10225 DATA"TILING & TERRAZZO",1.39,1.91,1.28,1.47
10230 DATA"SUSPENDED CEILINGS",1.13,2.43,.55,1.16
10235 DATA"GLASS",2.34,1,1.52,1.6
10240 DATA"DECORATIONS",2.45,1.29,3.4,2.64
10245 DATA"ELECTRICAL",8.42,8.1,9.52,8.89
10250 DATA"HEATING,VENTLN & AIR CONDTNG",11.31,11.31,13.7,12.5
10255 DATA"LIFT",0.82,3.85,0.45,1.39
10260 DATA"STRUCTURAL STEEL",12.99,0.44,0.44,3.58
```

```
   *** COST OF PLANNING A BUILDING ***
       ----------------------------
            THE B.C.I.S BUILDING
            --------------------
                COST INDICES
                ------------

OPTIONS AVAILABLE:-

<1>-ENTER BCIS INDICES.
<2>-PRINTOUT % WEIGHTINGS.
<3>-END PROGRAM.

INPUT THE NEDO INDEX

DEMOLITIONS                     256<
```

Screen display 3.1

```
BUILDING COSTS BASED ON
STEEL-FRAMED CONSTRUCTION.

    WORK       NEDO         %        EXTENSION
CATEGORY      INDEX    WEIGHTING

DEMOLITION     256         0.00           0.00

EXCAVATION     254         1.89         480.06

HARDCORE       255         1.18         300.90

GENERAL PI     262         0.00           0.00

SHEET STEE     216         0.00           0.00

CONCRETE       262         8.08        2116.96

    <PRESS 'SPACE BAR' TO CONTINUE>
```

Screen display 3.2

Chapter Four
Early Cost Estimating

Readers will no doubt be aware of the many and varied methods of estimating the cost of a building in the early design stages. These include simple single price estimates such as a cost per square metre through to elemental cost planning approaches and the measurement of simple approximate quantities for pricing. In our previous book 'Microcomputers in Building Appraisal', Granada 1983, we provided an elemental cost planning program which undertook the traditional adjustments for changes between a chosen cost analysis and a proposed project.

In this chapter we present an approach developed by Davis, Belfield and Everest, Chartered Quantity Surveyors, in 'The Architects' Journal' of 14 September 1977 and referred to here with their kind permission. Some alterations have been made to simplify the program but these are unlikely to make a major difference to the final result.

The method

The purpose of the method is to forecast the cost of a factory and associated offices knowing the specification of ten basic elements. The shape of the factory is assumed (these do not differ greatly in practice) and a scale is used to interpolate the cost between buildings of different size. The technique is explained in detail in 'The Architects' Journal' article and should be read by all serious users of the program but a brief resume is given below as a reminder.

At the heart of the method is a chart which shows the minimum and maximum cost per square metre for each element related to a scale which represents the floor area. This floor area extends from 500 to 5000 square metres and the data included will not work for buildings outside this range. The costs relate to various specifications and in the manual version the user was expected to select a cost within the range of the specification he had chosen.

A diagrammatic representation of this process is shown in Fig. 4.1 below:

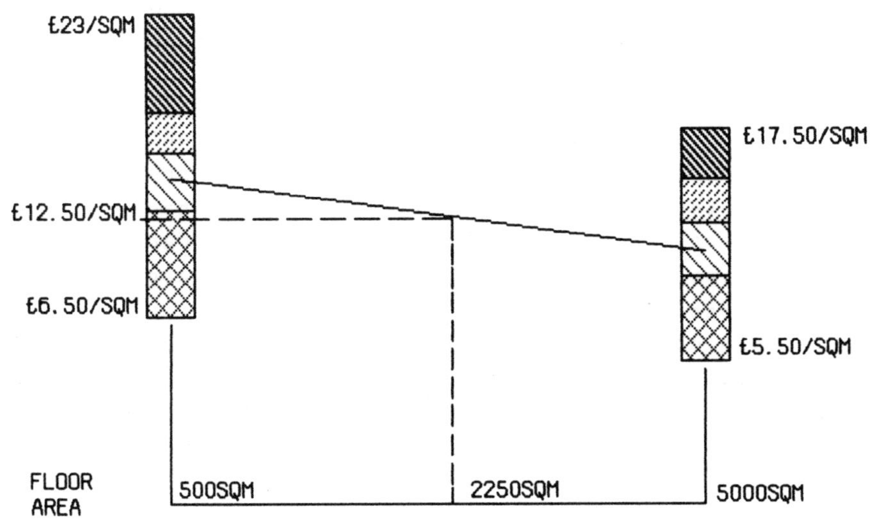

Figure 4.1 Interpolation of floor area to obtain cost per square metre

In this case specification B has been chosen for a building of 2250 square metres. It is merely a question of reading off the cost of B from the line on the graph (£12.50) and including this in the estimate for the element.

Two important points should be noted. Firstly in the manual version it was possible for the user to choose to join any two points in the specification 'boxes' thus giving some extra control as to the expected cost. In the computer program it is the approximate mid-point of the chosen specification 'box' which has been adopted. Secondly the scale along the bottom is not a natural scale and in fact varies slightly from element to element. This could be accommodated in the program but would lengthen it considerably. To simplify matters a log scale has been used which approximates to the original for the factory part of the estimate but may slightly distort the office calculation. As the method is only providing a 'ball park' estimate it was felt that this was not critical.

Once the cost per square metre has been established for each element based on the specification chosen, two other adjustments, for location and price level, need to be made. Fortunately Davis, Belfield and Everest have produced weightings for various regions of the country and the total cost per square metre is merely multiplied by this weighting to arrive at a reasonable cost for the geographical location.

Price levels are adjusted in the conventional way using 'The Architects' Journal' (AJ) tender based index. The total of the costs, adjusted for location, is multiplied by a ratio of the anticipated contract date AJ tender index to the September 1977 AJ tender index. Unfortunately since the article was published the base date for the tender index has altered and there is therefore an adjustment in line 1280 which allows the current tender index to be used.

A word of warning should be included here because both the location and price level indices are rather blunt instruments. The skill and judgement of the cost adviser should be used to assess the particular needs of a specific site and building market if a realistic figure is to be obtained.

The procedure is identical for both factory and offices and indeed the final adjustments need not be made until both specification selections have been made. There are however one or two points to note with regard to the offices.

Three main office categories have been identified:

1. Offices within the main structure at ground floor level.
2. Offices within the main structure at a higher (first floor) level.
3. Offices outside the structure but attached to an external wall.

In 1. the offices are costed as 'extra over' basic costs and effectively no additional area is added to the overall area of the building.
In 2. the costs are again extra over but in this case additional area is added at first floor level.
In 3. the costs are full (total) prices and the additional area is added to that already included in the main structure.

The data for the offices has been based on a figure of around ten per cent of the main workshop/factory area but can be used for any office area required within five to twenty per cent of the gross floor area.

The results

The results from the above calculations are computed as both a total figure and as a cost per square metre for factory and office. It is then possible to request an elemental breakdown of the figures and this will be shown together with the options chosen. The AJ article contained a number of additional qualifications for the use of the method and it is advisable to consult this before use.

The following inputs should provide the results shown in Table 4.1.

Factory	Area of factory	3000 square metres
	Location	Outer London
	AJ cost index	225
	Specification selection no. 2 throughout except sanitary accommodation.	
Offices	Offices type	External 300 sq.m
	Specification selection	Foundations no. 4

Frames	no. 3
Upper floor	no. 1
Roofs	no. 5
External walls	no. 3
Extl windows	no. 2
Partitions	no. 2
Finishes	no. 2
Sanitary ftgs	no. 4
Elec services	no. 2

Table 4.1 Results

FACTORIES PROGRAM

	COST/M2	TOTAL COST
FACTORY	164.58	493740
OFFICES	314.97	94491
TOTAL	178.25	588231

BREAKDOWN OF THE FACTORY/OFFICE COSTS

ELEMENT	FACTORY	OFFICES
FOUNDATIONS & FL	32.6	34.19
FRAMES	31.98	66.94
ROOFS & COVERING	26.17	38.67
UPPER FLOORS	0	0
EXTERNAL WALLS	20.2	20.3
EXTERNAL WINDOWS	10.14	15.43
INTERNAL DIVISIO	0	28.27
FINISHES	7.95	30.36
SANITARY ACCOMMO	7.25	0
MECHANICAL SERVI	16.57	0
SANIT'Y PLUMB. &	0	39.61
ELECTRICAL SERVI	7.32	41.2
FIRE PROTECTION	4.4	0
TOTAL	164.58	314.97

Description of Program 4

Line 1020	Dimensions the necessary arrays. A$(),A(),B() being the arrays used for storing the different specifications of each element. A$() is the specification description. A() is the related constant of the line. B() is the related gradient of the line. E$() stores the element titles. P() stores the element costs for the factory. Q() stores the element costs for the office.
Lines 1025-1035	Initialise the general variables to be used in the program (see Chapter 11). Variable LL enables the user to specify the screen width to be used. (Set variable to one less than the number of screen columns.)
Lines 1040-1090	Display some lines of explanation as to the purpose of the program.
Lines 1095-1130	Ask whether the results are to be displayed on the screen or sent to the printer. P$="1" for screen display P$="2" for printer.
Lines 1135-1215	Ask for the input of the ground floor area of the factory. (Must be between 500 and 5000 sq.m otherwise the input is ignored.) The area is 'scaled' by taking its logarithm at line 1275 and placing it into variable X.
Lines 1220-1235	Select from a list of possible geographical areas the location factor for the project.
Lines 1240-1270	Ask for the input of 'The Architects' Journal' tender price

	index for the proposed factory.
Line 1275	Scales factory area by taking its logarithm.
Line 1280	Amends the T.P.I. for the change in base date (see previously).
Line 1285	These two index factors are then combined into one.
Lines 1290-1330	Input and calculation of the factory specification cost. Line 1310 sets up the loop for the ten elements, the specification selection is made at subroutine 1700 (see later) which returns the description in A$(QQ), the gradient as B(QQ) and the constant as A(QQ). The results of the calculation are placed in array P(). Note that the element number is defined by variable 'N' not 'T'. (see later).
Lines 1335-1455	Deal with the office specification, if any. If an associated office is to be considered then its area and whether it is internal or external to the factory need to be stated. As with the factory the area is scaled logarithmically at line 1385. Input of the office specification is made at lines 1445 - 1455, the results being placed into array Q(). It is worthwhile reiterating that the offices are dealt with on an 'extra over' basis.
Lines 1460-1560	Display/Print the total costs for the factory and its office (these are calculated at lines 1485 to 1495) using subroutine 8500 to format the output correctly.
	The summations of arrays P() and Q() are placed in variables J1 and J2 respectively and are then rounded off. The loop at line 1485 is set from 1 to 20 as not all elements are

	applicable to both the factory and the office. (See 'subroutines' below.) These variables represent the total costs per square metre which are then multiplied by their respective areas to find the total cost, which is placed in variable 'L'. Depending on whether the office, if any, is external or internal the average cost per square metre is found by dividing the sum of the two areas or only the factory area (line 1495).
Lines 1565-1665	Display/Print the breakdown of the element costs if required, otherwise the program jumps to the 'end of program' routine.

<u>Subroutines</u>

Line 1675	Clears the screen and takes the cursor down five lines.
Line 1680	Halts the program until 'space' key is pressed and then calls subroutine 1675.
Line 1685	Opens the printer channel - turns printer on.
Line 1690	Closes the printer channel - turns printer off.
Lines 1700-1770	Read and then display the specification ready for selection. An explanation of how the data is ordered is is required here: In line 1700, 'N' is the number of the element in the array of elements E$(). This allows the office to have specific elements that are not directly applicable to a factory, for example internal divisions. Similarly mechanical services are not applicable to offices.

E$ is a temporary store for the element name. ZZ is the number of different specifications available.
Lines 1710-1735 display the possible specifications formatted to the screen width specified by the user. (See line 1035.) If the specification description is longer than the width of the screen it is neatly divided in two and then displayed.
Line 1730 is merely a time delay to enable the operator to see the input selection, if using the screen only.

<u>List of elements</u>

1. Foundations and floor slab.
2. Frame.
3. Roof coverings.
4. Upper floors (offices only).
5. External walls.
6. External windows and doors.
7. Internal divisions (offices only).
8. Finishes.
9. Sanitary accommodation (factory only).
10. Mechanical services.
11. Sanitary, plumbing and mechanical services (offices only).
12. Electrical services.
13. Fire protection (factory only).

Each specification has a description and two values associated with it which describe the straight line taken from the Davis, Belfield and Everest charts. The first is the line constant, (the theoretical cost when the area used is zero) followed by the gradient (see below).

Cost/sq.m = constant - gradient * area (scaled)
$$y = a - bx$$

The line always has a negative gradient explained by the costs per square metre diminishing as the building size increases.

Postscript on printers

(1) The option to select a hardcopy of the results (lines 1500-1665) may not be possible on some computers without some amendments to the coding. Specifically, the program can be used on: other Commodore, BBC and Apple computers with some changes to subroutines 1685 and 1690, these being:

```
COMMODORE    As listing

BBC          1685  *FX6,0
             1686  VDU 2:VDU 21:Z$="":RETURN
             1690  VDU 1,10:VDU 6:VDU 3:RETURN

APPLE        1685  PRINT CHR$(4);"PR#1":Z$=LF$:RETURN
             1690  PRINT:PRINT CHR$(4);"PR#0":RETURN
```

Computers using Microsoft BASIC will have to use a duplicate set of program statements for lines 1500-1665 replacing all PRINT statements with LPRINT. Similarly the PRINT statements at lines 1195-1215, 1270, 1435-1440 and 1770 will have to be replaced with LPRINT. Further necessary line amendments are as follows:

```
1685  Z$=LF$:RETURN
1690  RETURN
```

(2) The variable LF$ may be set to a line-feed control character (CHR$(10)) if the printer cannot accommodate an automatic line-feed. (See also Chapter 11.)

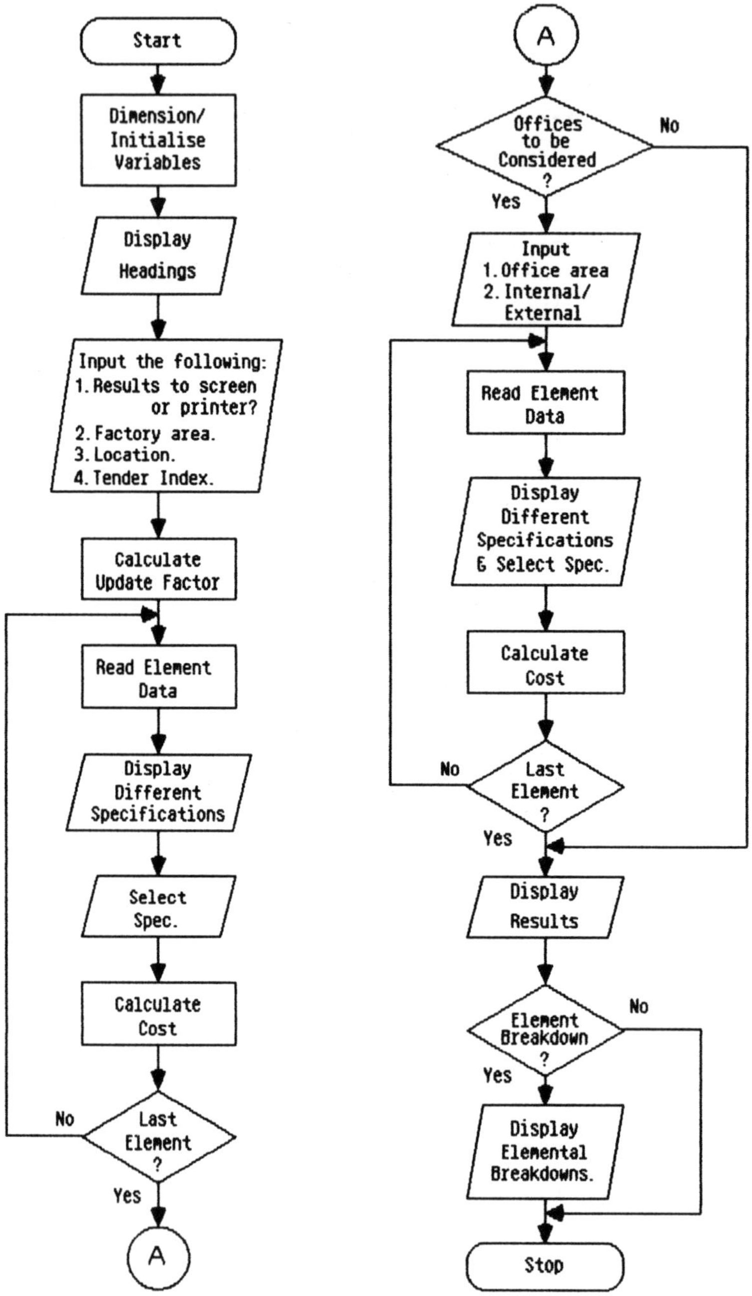

Figure 4.2 Flowchart for Program 4: Early cost estimating for factories

```
1000 REM *********** EARLY COST ESTIMATING. PROGRAM 4. ************

1005 REM ************ COPYRIGHT BRANDON, MOORE & MAIN *************
1010 REM *COLLINS PROFESSIONAL & TECHNICAL BOOKS (C)85 REV 24/1/85*

1015 REM ************ DIMENSION/INITIALISE VARIABLES **************
1020 DIM A$(10),A(10),B(10),E$(20),P(20),Q(20)
1025 SP$="":LT$="":FOR Z= 1 TO 79:SP$=SP$+" ":LT$=LT$+"-":NEXT Z
1030 DEF FNA(X)=INT(.5+100*X)/100
1035 CS$=CHR$(147):HS$=CHR$(19):RM$="↑":LF$="":CL=0:OT$="":LL=40

1040 REM ****************** DISPLAY HEADINGS *********************
1045 GOSUB 1675:PRINT"THIS PROGRAM IS BASED ON":PRINT
1050 PRINT"THE INITIAL COST ESTIMATING":PRINT
1055 PRINT"GUIDES PUBLISHED IN THE":PRINT
1060 PRINT"ARCHITECTS JOURNAL & PREPARED":PRINT
1065 PRINT"BY DAVIS, BELFIELD & EVEREST":PRINT
1070 PRINT"CHARTERED QUANTITY SURVEYORS."
1075 GOSUB 1680:PRINT"THIS PROGRAM ATTEMPTS TO GIVE":PRINT
1080 PRINT"EARLY COST ADVICE TO CLIENTS":PRINT
1085 PRINT"WISHING TO BUILD FACTORIES":PRINT
1090 PRINT"WITH ASSOCIATED OFFICES...."

1095 REM ************** OUTPUT TO SCREEN OR PRINTER ***************
1100 GOSUB 1680:PRINT"DO YOU HAVE A PRINTER ATTACHED?":PRINT
1105 PRINT"IF SO DO YOU REQUIRE A HARD COPY":PRINT
1110 PRINT"OF THE SELECTIONS/RESULTS?":PRINT
1115 PRINT"ANSWER:- <1> FOR SCREEN ONLY.":PRINT
1120 PRINT"          <2> FOR PRINTOUT.":PRINT
1125 GOSUB 5500:P$=Z$:IF P$<>"1" AND P$<>"2" THEN 1125
1130 PRINT"OPTION ";P$;" SELECTED.":GOSUB 1680

1135 REM ****************** INPUT FACTORY AREA *******************
1140 PRINT"PLEASE STATE AREA OF":PRINT
1145 PRINT"FACTORY TO BE CONSIDERED":PRINT
1150 PRINT"INCLUDING ANY INTERNAL":PRINT
1155 PRINT"GROUND FLOOR OFFICES.":PRINT
1160 PRINT"(500-5000M2 ONLY)":PRINT:PRINT:PRINT
1165 PRINT"N.B. EXTERNAL OFFICES & "
1170 PRINT"FIRST FLOOR OFFICES WILL"
1175 PRINT"BE CONSIDERED LATER."
1180 RN=14:GOSUB 5000:INPUT A
1185 IF A<500 OR A>5000 THEN GOSUB 1675:GOTO 1140
1190 IF P$="1" THEN 1225
1195 GOSUB 1685:PRINT TAB(11)"FACTORIES PROGRAM.";Z$
1200 PRINT TAB(11);LEFT$(LT$,18);Z$:PRINT Z$
1205 PRINT"AREA OF FACTORY (INC.INTERNAL GRD FLR OFFICES)=";A;"M2";Z$
1210 PRINT:PRINT"THE FOLLOWING SELECTIONS WERE MADE:";Z$
```

```
1215 PRINT LEFT$(LT$,35);Z$:GOSUB 1690


1220 REM **************** INPUT LOCATION FACTOR *******************
1225 GOSUB 4000:PRINT"IN WHICH OF THE FOLLOWING":PRINT LEFT$(LT$,25)
1230 PRINT"LOCATIONS IS YOUR SITE SITUATED":PRINT LEFT$(LT$,32)
1235 T=0:GOSUB 1700:C=A(QQ)


1240 REM **************** INPUT A.J. TENDER INDEX *****************
1245 GOSUB 1680:PRINT"WHAT IS THE 'A.J.' CONTRACT":PRINT
1250 PRINT"DATE TENDER INDEX FOR THE":PRINT
1255 PRINT"CONTRACT BEING CONSIDERED.":PRINT:PRINT
1260 INPUT"ENTER APPROPRIATE NUMBER ";T
1265 IF P$="1" THEN 1275
1270 GOSUB 1685:PRINT "CONTRACT DATE TENDER INDEX=";T;Z$:GOSUB 1690
1275 X=LOG(A):REM***SCALE FACTORY AREA TO LOG SCALE
1280 T1=(1+(263-233)/233)*100:REM***AMEND INDEX TO UPDATE CORRECTLY
1285 F=C*(T/T1):REM***COST FACTOR FOR UPDATE TO TPI & LOCATION


1290 REM ************** INPUT FACTORY SPECIFICATION ***************
1295 GOSUB 1680:PRINT"ANSWER THE FOLLOWING QUESTIONS":PRINT
1300 PRINT"FOR THE FACTORY SPECIFICATION":PRINT
1305 PRINT"WITH A NUMBER ONLY.":GOSUB 1680
1310 FOR T= 1 TO 10
1315 GOSUB 1675:GOSUB 1700:P(N)=(A(QQ)-(B(QQ)*X))*F
1320 P(N)=FNA(P(N)):NEXT T
1325 GOSUB 1675:PRINT"THIS COMPLETES THE FACTORY SPECIFICATION"
1330 PRINT LEFT$(LT$,40):PRINT:PRINT


1335 REM *************** INPUT OFFICE SPECIFICATION ***************
1340 PRINT"DO YOU REQUIRE THE ADDITIONAL COST"
1345 PRINT"OF THE OFFICES? (Y/N)"
1350 GOSUB 5500:A$=Z$:IF A$<>"Y" AND A$<>"N" THEN 1350
1355 IF A$="N" THEN E=0:GOTO 1465
1360 GOSUB 1675:PRINT"PLEASE STATE THE AREA OF THE":PRINT
1365 PRINT"OFFICES TO BE CONSIDERED IN M2.":PRINT
1370 PRINT"(50-500 M2)":PRINT
1375 INPUT"ENTER APPROPRIATE FIGURE ";E
1380 IF E<50 OR E>500 THEN 1360
1385 Y=LOG(E):PRINT:PRINT
1390 PRINT"IS YOUR OFFICE GROUND FLOOR INTERNAL? (Y/N)"
1395 GOSUB 5500:A$=Z$:IF A$<>"Y" AND A$<>"N" THEN 1395
1400 GOSUB 1675:PRINT"YOU WILL NOW BE ASKED TO SELECT THE":PRINT
1405 PRINT"SPECIFICATION FOR THE OFFICES ONLY.":PRINT:PRINT
1410 PRINT"N.B. ONLY SPECIFY THE ADDITIONAL WORK"
1415 PRINT"REQUIRED OVER & ABOVE THAT PREVIOUSLY"
1420 PRINT"SPECIFIED FOR THE FACTORY.":GOSUB 1680
1425 IF P$="1" THEN 1445
1430 GOSUB 1685:OT$="INTERNAL":IF A$="N" THEN OT$="EXTERNAL"
1435 PRINT Z$:PRINT"OFFICE SPECIFICATION.";Z$:PRINT LEFT$(LT$,21);Z$
1440 PRINT Z$:PRINT OT$;" OFFICE AREA =";E;" M2";Z$:GOSUB 1690
1445 FOR T= 1 TO 10
```

```
1450 GOSUB 1675:GOSUB 1700:Q(N)=(A(QQ)-(B(QQ)*Y))*F
1455 Q(N)=FNA(Q(N)):NEXT T


1460 REM **************** PRINT OUT RESULT TABLE ******************
1465 GOSUB 1675:PRINT"THIS COMPLETES THE OFFICE SPECIFICATION"
1470 PRINT:PRINT"THE RESULTS WILL BE TYPED OUT ON THE"
1475 IF P$="1" THEN PRINT"SCREEN.":Z$=""
1480 IF P$="2" THEN PRINT"PRINTER.":GOSUB 1685:PRINT Z$
1485 J1=0:J2=0:FOR T= 1 TO 20:J1=J1+P(T):J2=J2+Q(T):NEXT T
1490 J1=FNA(J1):J2=FNA(J2):K1=J1*A:K2=J2*E:L=K1+K2
1495 T1=L/(A+E):IF OT$="INTERNAL" THEN T1=L/A
1500 PRINT"FACTORIES PROGRAM.";Z$
1505 PRINT LEFT$(LT$,18);Z$:PRINT Z$
1510 PRINT LEFT$(SP$,13);"COST/M2        TOTAL COST";Z$
1515 PRINT LEFT$(SP$,13);"-------        ----------";Z$:PRINT Z$
1520 B=J1:GOSUB 8500
1525 PRINT"FACTORY    ";N$;"        ";RIGHT$(SP$+STR$(INT(K1)),10);Z$
1530 B=J2:GOSUB 8500
1535 PRINT"OFFICES    ";N$;"        ";RIGHT$(SP$+STR$(INT(K2)),10);Z$
1540 PRINT"                ";LEFT$(LT$,26);Z$
1545 B=T1:GOSUB 8500
1550 PRINT"TOTAL      ";N$;"        ";RIGHT$(SP$+STR$(INT(L)),10);Z$
1555 PRINT"                ";LEFT$(LT$,26);Z$:PRINT Z$
1560 IF P$="2" THEN GOSUB 1690:PRINT


1565 REM ************ PRINT ELEMENTAL BREAKDOWN TABLE *************
1570 PRINT"DO YOU REQUIRE AN ELEMENTAL BREAKDOWN?"
1575 PRINT"ENTER <Y>ES OR <N>O."
1580 GOSUB 5500:A$=Z$:IF A$<>"Y" AND A$<>"N" THEN 1580
1585 IF A$="N" THEN 1660
1590 IF P$="1" THEN GOSUB 4000:Z$="":REM              CLEARS SCREEN
1595 IF P$="2" THEN GOSUB 1685:PRINT Z$:PRINT Z$
1600 PRINT"BREAKDOWN OF THE FACTORY/OFFICE COSTS";Z$
1605 PRINT LEFT$(LT$,38);Z$:PRINT Z$
1610 PRINT"     ELEMENT           FACTORY    OFFICES";Z$
1615 PRINT"     -------           -------    -------";Z$
1620 T1=0:T2=0:FOR T= 1 TO 20:IF E$(T)="" THEN 1635
1625 B=P(T):GOSUB 8500:M$=N$:B=Q(T):GOSUB 8500
1630 PRINT LEFT$(E$(T)+SP$,18);" ";M$;"  ";N$;Z$
1635 NEXT T
1640 PRINT"             ";LEFT$(LT$,30);Z$
1645 B=J1:GOSUB 8500:M$=N$:B=J2:GOSUB 8500
1650 PRINT"TOTALS                  ";M$;"  ";N$;Z$
1655 PRINT"             ";LEFT$(LT$,30);Z$
1660 IF P$="2" THEN GOSUB 1690
1665 GOSUB 1680:GOTO 9998:REM*** END OF PROGRAM


1670 REM ****************** PROGRAM SUBROUTINES ******************
1675 GOSUB 4000:RN=5:GOSUB 5000:RETURN:REM***CLEAR SCREEN & GO DOWN
1680 GOSUB 9000:GOSUB 1675:RETURN:REM      *** PAUSE & CLEAR SCREEN
1685 OPEN2,4:CMD2:Z$=LF$:RETURN:REM        *** OPEN  PRINTER     !!!
1690 PRINT#2:CLOSE2:RETURN:REM              *** CLOSE PRINTER     !!!
```

```
1695 REM ******* READ/DISPLAY DATA AND SELECT SPECIFICATION *******
1700 READ N,E$,ZZ:E$(N)=E$:PRINT:PRINT:PRINT"ELEMENT:- ";E$:PRINT
1705 FOR QQ= 1 TO ZZ:READ A$(QQ),A(QQ),B(QQ)
1710 IF LEN(A$(QQ))<LL THEN PRINT A$(QQ):GOTO 1740
1715 FOR SS= LL TO 1 STEP(-1)
1720 IF RIGHT$(LEFT$(A$(QQ),SS),1)=" " THEN PP=SS-1:SS=1
1725 NEXT SS
1730 PRINT LEFT$(A$(QQ),PP)
1735 PRINT "  ";RIGHT$(A$(QQ),LEN(A$(QQ))-PP)
1740 NEXT QQ
1745 RN=23:GOSUB 5000:PRINT"ENTER APPROPRIATE NUMBER."
1750 GOSUB 5500:A$=Z$:QQ=VAL(A$):IF QQ=0 OR QQ>ZZ THEN 1750
1755 RN=23:GOSUB 5000:PRINT A$;" SELECTED.                 "
1760 IF P$="2" THEN 1770
1765 FOR Z= 1 TO 500:NEXT Z:RETURN
1770 GOSUB 1685:PRINT E$;A$(QQ);Z$:GOSUB 1690:RETURN
```

The following subroutines are required. (See Chapter 11)

4000 Clear screen routine.
4500 Home screen routine.
5000 Cursor position routine.
5500 Input character routine.
8500 Format number to two decimal places.
9000 Pause routine.
9998 End program routine.

```
10000 REM ****************** SPECIFICATION DATA ******************
10005 DATA0,"LOCATION                          ",9
10010 DATA" 1-OUTER LONDON.",1,0," 2-INNER LONDON.",1.08,0
10015 DATA" 3-LIVERPOOL.",1.05,0," 4-SOUTH EAST.",0.95,0
10020 DATA" 5-SCOTLAND & WALES.",0.93,0
10025 DATA" 6-EAST/WEST MIDLANDS.",0.92,0
10030 DATA" 7-EAST ANGLIA.",.91,0
10035 DATA" 8-NORTHWEST OR YORKSHIRE/HUMBERSIDE.",0.89,0
10040 DATA" 9-NORTHERN OR SOUTHWEST.",0.88,0
10045 DATA1,"FOUNDATIONS & FLR SLAB    ",4
10050 DATA" 1-AVERAGE SLAB & BASE.",28.207,1.945
10055 DATA" 2-DEEPER TRADITIONAL FOUNDATIONS.",35.332,2.37
10060 DATA" 3-RAFT FOUNDATIONS.",47.917,3.363
10065 DATA" 4-PILES.",68.765,5.214
10070 DATA2,"FRAME                             ",5
10075 DATA" 1-STEEL PORTAL.",31.523,2.316
10080 DATA" 2-PRECAST CONCRETE PORTAL.",41.073,3.126
10085 DATA" 3-STEEL PORTAL,ENCASED COLNS/BEAMS",47.777,3.654
10090 DATA" 4-NORTH LIGHT FRAME.",55.032,4.156
10095 DATA" 5-SPACE DECK.",63.178,4.597
10100 DATA3,"ROOF COVERINGS                    ",4
10105 DATA" 1-INS.ASBESTOS(5% TRANSLUCENT SHEETS)",19.65,1.239
10110 DATA" 2-FLAT METAL DECK(RF FELT & CHIP'GS)",25.040,1.488
10115 DATA" 3-COMPOSITE INSULATED POLYESTER ON METAL DECK",33.524,2.0
      06
```

```
10120 DATA" 4-NORTH LIGHT.(COL STEEL INNER LINING & PAT.GLAZ'G)"
10125 DATA49.476,3.186
10130 DATA5,"EXTERNAL WALLS                  ",5
10135 DATA" 1-INSULATED ASBESTOS.",41.165,4.278
10140 DATA" 2-PVC COATED STEEL CLADDING WITH INNER LINING",44.951,4.3
      49
10145 DATA" 3-FACED BRICK & BLOCK CAVITY WALLS",60.742,6.193
10150 DATA" 4-PRECAST CONCRETE PANELS",80.537,8.136
10155 DATA" 5-INSULATED G.R.P. SANDWICH",117.616,11.616
10160 DATA6,"EXTERNAL WINDOWS & DOORS ",4
10165 DATA" 1-MANUAL DOORS/TIMB.WINDOWS-6% W.AREA",31.181,3.45
10170 DATA" 2-ELECT.DOORS/METAL WINDOWS-6% W.AREA",43.575,4.807
10175 DATA" 3-MANUAL DOORS/METAL WIND'S-20% W.AREA",57.101,6.286
10180 DATA" 4-ELECT.DOORS/METAL WINDOWS-20% W.AREA",73.185,8.05
10185 DATA8,"FINISHES                        ",3
10190 DATA" 1-MIN. DECORATION POWER FLOATED SLAB.",8.733,0.782
10195 DATA" 2-STEELWORK PAINTED,EARLY GRINDING SLAB FINISH.",10.953,.
      87
10200 DATA" 3-INT.PAINTING,GRANOLITHIC FLR FINISH.",14.49,0.871
10205 DATA9,"SANITARY ACCOMODATION     ",1
10210 DATA" 1-TOILETS FOR FACTORY & INT.DRAINAGE.",18.289,1.83
10215 DATA10,"MECHANICAL SERVICES       ",5
10220 DATA" 1-NONE REQUIRED.",0,0
10225 DATA" 2-HOT AIR SPACE HEATING,GAS FIRED.",16.688,1.046
10230 DATA" 3-LOW PRESSURE HOT WATER UNIT.",23.069,1.4
10235 DATA" 4-RADIANT HEATING.",27.524,1.697
10240 DATA" 5-HEATING AND VENTILATION.",31.009,1.716
10245 DATA12,"ELECTRICAL SERVICES       ",3
10250 DATA" 1-MAIN SWITCH GEAR ONLY.",10.547,1.107
10255 DATA" 2-SERVICES TO TOILETS & MECH.SERV.ONLY",14.665,1.373
10260 DATA" 3-LIGHTING & POWER TO FACTORY.",23.135,1.803
10265 DATA13,"FIRE PROTECTION           ",3
10270 DATA" 1-HOSE REELS OR FIRE ALARMS.",5.527,0.518
10275 DATA" 2-HOSE REELS & FIRE ALARMS.",8.821,0.826
10280 DATA" 3-HOSE REELS & FIRE ALARMS + SMOKE/HEAT DETECTORS."
10285 DATA11.805,1.002
10290 DATA1,"FOUNDATIONS & FLOOR SLAB ",5
10295 DATA" 1-NONE.",0,0
10300 DATA" 2-INTERNAL ON RAFT OR SLAB & BASES.",4.92,0.554
10305 DATA" 3-FIRST FLOOR ON SLAB & BASES.",20.668,1.886
10310 DATA" 4-EXTERNAL ON SLAB & BASES.",30.769,2.387
10315 DATA" 5-FIRST FLOOR COLUMNS OFF PILES.",55.373,5.245
10320 DATA2,"FRAMES                          ",4
10325 DATA" 1-NONE.",0,0
10330 DATA" 2-STEEL FRAME TO 1ST FLOOR OFFICES.",34.109,2.036
10335 DATA" 3-STEEL FRAME & LTD.PROTCTN.INT.LOAD CONC RF.OVER GRD.FLR
      .OFFICES"
10340 DATA58.314,4.336
10345 DATA" 4-STEEL FRAME & PROTECTION TO 2 STOREY OFFICES"
10350 DATA70.234,5.612
10355 DATA4,"UPPER FLOORS                    ",2
10360 DATA" 1-NONE",0,0
10365 DATA" 2-TO 1ST FLOOR OFFICES",29.851,2.696
10370 DATA3,"ROOFS & COVERINGS          ",6
10375 DATA" 1-NONE.",0,0
```

```
10380 DATA" 2-FIRE PROTECTION & FRAMING TO 1ST FLR",10.037,0.934
10385 DATA" 3-JOIST & CHIPBOARD RF.TO INT.OFFICES",17.08,1.311
10390 DATA" 4-SIMPLE FELT COVERED RF.TO EXT.OFFICES",24.337,1.681
10395 DATA" 5-ASPHALT RF.TO EXT.OFFICES WITH BRICK PARAPETS & GUTTERS
      "
10400 DATA28.319,1.564
10405 DATA" 6-EXT. ROOF WITH UPTO 10% ROOFLIGHTS",33.836,1.653
10410 DATA5,"EXTERNAL WALLS              ",3
10415 DATA" 1-NONE.",0,0
10420 DATA" 2-BLOCKWK INNER LIN'G TO 1ST FLR OFFCS",10.27,1.357
10425 DATA" 3-FACED BRKWK & BLKWK CAVITY WALLS TO EXT. OFFICES",36.42
      2,4.6
10430 DATA6,"EXTERNAL WINDOWS & DOORS ",3
10435 DATA" 1-NONE.",0,0
10440 DATA" 2-WINDOWS & DOORS TO EXT.OFFICES.",26.233,3.242
10445 DATA" 3-WNDWS & DRS TO 1ST FLR & EXT.OFFICES",36.908,4.363
10450 DATA7,"INTERNAL DIVISIONS & DRS ",2
10455 DATA" 1-LTD PARTITIONS TO OFFICES.",34.533,4.538
10460 DATA" 2-FULLY PARTITIONED AREAS TO OFFICES.",47.148,5.78
10465 DATA8,"FINISHES                   ",3
10470 DATA" 1-MIN.FINISHES TO GRD FLR OFF.(PLASTER UNDERCT,FLR SCRD &
       ARTEX)"
10475 DATA26.195,2.828
10480 DATA" 2-MIN.FINISHES TO 1ST FLR OR AVER.",35.991,3.64
10485 DATA" 3-AVERAGE FINISHES TO 1ST FLR.OFFICES",48.693,4.918
10490 DATA11,"SANIT'Y PLUMB.& MECH.SERV",5
10495 DATA" 1-SANITARY FITTINGS TO GRD.FLR.OFFICES",22.67,3.154
10500 DATA" 2-SANITARY FITTINGS TO 1ST.FLR.OFFICES",30.432,4.036
10505 DATA" 3-SANITARY FITTINGS H&C WATER SERVICES",45.295,5.448
10510 DATA" 4-RADIATOR HEATING TO OFFICES",52.955,5.8
10515 DATA" 5-SAN.FITTGS H&C WATER RAD.HTG.TO OFF.",69.458,6.009
10520 DATA12,"ELECTRICAL SERVICES        ",2
10525 DATA" 1-GEN.LIGHTING & POWER TO SPECULATIVE OFFICES",25.182,1.8
      41
10530 DATA" 2-HIGHER STANDARD TO OWNER OCCP OFFCS",36.131,2.711
```

Screen display 4.1

```
IN WHICH OF THE FOLLOWING
-------------------------
LOCATIONS IS YOUR SITE SITUATED
--------------------------------


ELEMENT:- LOCATION

 1-OUTER LONDON.
 2-INNER LONDON.
 3-LIVERPOOL.
 4-SOUTH EAST.
 5-SCOTLAND & WALES.
 6-EAST/WEST MIDLANDS.
 7-EAST ANGLIA.
 8-NORTHWEST OR YORKSHIRE/HUMBERSIDE.
 9-NORTHERN OR SOUTHWEST.




ENTER APPROPRIATE NUMBER.
```

Screen display 4.2

```
ELEMENT:- FOUNDATIONS & FLR SLAB

 1-AVERAGE SLAB & BASE.
 2-DEEPER TRADITIONAL FOUNDATIONS.
 3-RAFT FOUNDATIONS.
 4-PILES.




ENTER APPROPRIATE NUMBER.
```

Chapter Five
Regression Analysis

Data based on an activity such as building, which depends to a great extent on human performance, will not give exact relationships between variables. To take a simple example, if we are trying to estimate the cost of brickwork we need to know how many bricks per hour will be laid by each tradesman. An experiment to determine this value using twenty different bricklayers will probably give twenty different answers. All we can discover from such an experiment is an average value.

Where a number of variables are involved such as the shape, height, finishes, etc., of a particular building type we need to be able to determine a trend algebraically in order to produce useful information for costing. A method of doing this is by the use of a regression model which is a means of determining the relationship between a number of variables by the 'method of least squares'. This method seeks to minimise the sum of the squares between the observed and predicted values.

To illustrate the principle on which the program given later in this chapter is based, we will take a simple case involving two variables - the height of a building and its cost per square metre of gross floor area. If a number of buildings are analysed we can obtain a set of values which may be plotted in the form of a graph. If we can then draw a straight line which reasonably fits the

---

For further reference on this subject see 'Cost Planning of Buildings' (5th edition) Ferry, D.J. and Brandon, P.S., Granada 1984. Chapter 15.

points on the graph we have a 'two - variable' linear regression. Obviously there are many situations which do not provide a linear relationship, in which case other methods of analysis would be used.

The factor whose value we wish to estimate (e.g. cost) is referred to as the 'dependent variable' and is plotted on the y axis. The factor from which these estimates are made is called the 'independent variable' and is plotted on the x axis. (Cause and effect are not necessarily implied by the terms 'dependent' and 'independent'.)

In two-variable linear regression the expression for the straight line is of the form:

$$y = a + bx$$

where a is the intercept on the y axis and b is the slope of the line (Fig. 5.1).

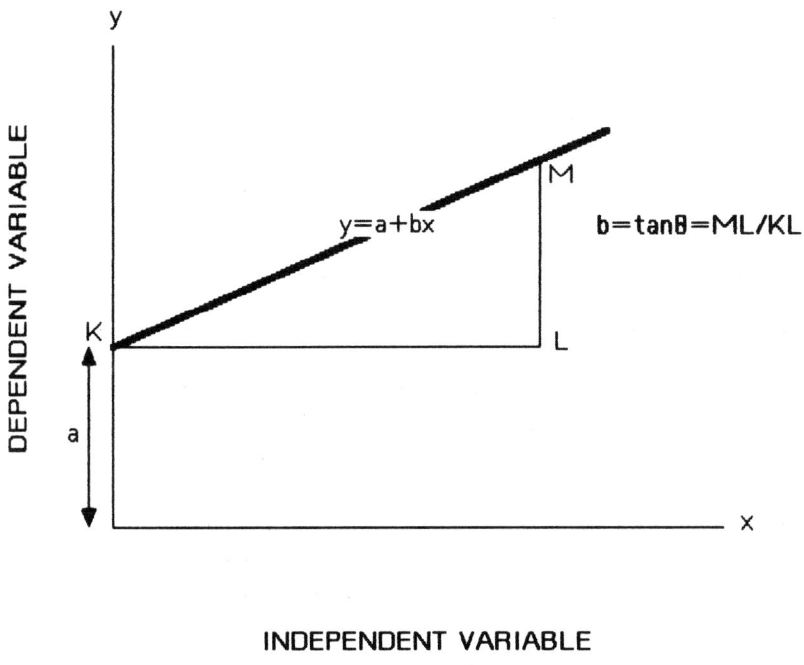

Figure 5.1 Formula for a straight line

To illustrate the technique, a hypothetical example has been chosen. It is assumed that there is a relationship between building height and building cost per square metre of floor area. The first step is to ascertain whether or not a linear relationship exists. In a simple case such as this we can plot the values to allow a visual examination. For the sake of simplicity, data from ten buildings of similar function and construction have been obtained and summarised in the table below. Such a small number would be an inadequate sample in practice - at least thirty data points should be found to give a reliable result.

Table 5.1 Hypothetical example: Building cost related to height

Building	Height (m)	Cost/sq. m (gross floor area)
A	4	500
B	20	544
C	17	580
D	10	570
E	7	504
F	23	560
G	24	600
H	6	524
I	10	530
J	15	542

These data are now plotted on a graph to produce what is known as a scatter diagram (Fig. 5.2). From this diagram it can be seen that there is a trend for costs to increase with height and that this trend appears to be linear rather than curved or heavily dispersed.

What we now need to find is a line which will fit these data in such a way that it will provide a reliable estimate of cost related to building height. This line of best fit is established by the technique of 'least squares'. The method requires that the sum of the squares of the deviations of the observed values of the dependent variable from the corresponding values on the chosen regression line must be a minimum. Figure 5.3 shows the principles of this technique in diagrammatic form using

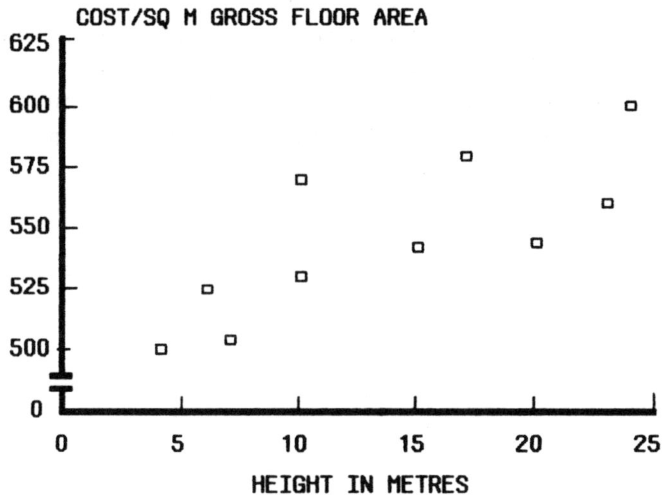

Figure 5.2 Scatter diagram

only four data points. A number of different lines can be drawn through these points and also through the average values of x and y (which is a requirement of this technique). Of the three lines drawn it is clear that position three provides the least squared deviation from the line. By the use of calculus to apply the condition that the sum of the squares of the deviations from a straight line must be a minimum, two equations result, one for 'a' and one for 'b' in the straight line formula.

$$b = \frac{\Sigma xy - n\bar{x}\bar{y}}{\Sigma x^2 - n\bar{x}^2}$$

$$a = \bar{y} - b\bar{x}$$

where $\bar{x}$ and $\bar{y}$ are the arithmetic means of the x and y variables and n is the number in the sample.

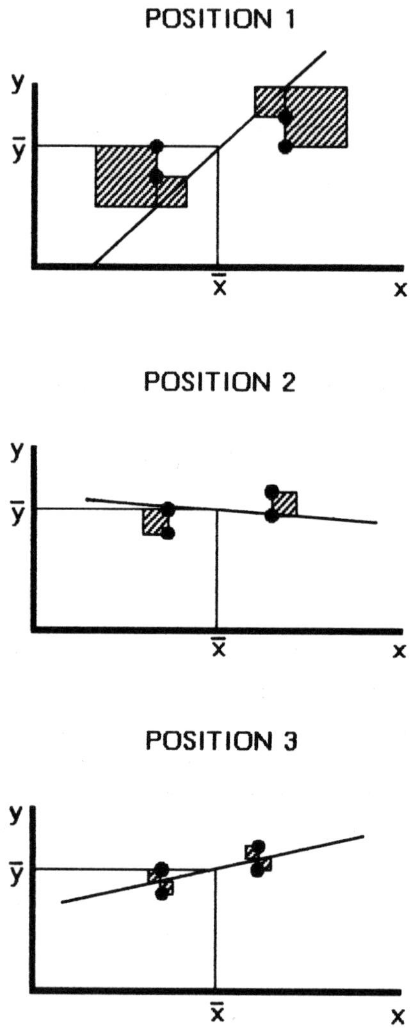

Figure 5.3 The principles of least squares. (Note: The smaller squared deviations from the regression line in position 3 makes this the line of best fit)

Using the data in table 5.1 the equations can be solved:

Building	x	y	xy	x^2	y^2
A	4	500	2000	16	250000
B	20	544	10880	400	295936
C	17	580	9860	289	336400
D	10	570	5700	100	324900
E	7	504	3528	49	254016
F	23	560	12880	529	313600
G	24	600	14400	576	360000
H	6	524	3144	36	274576
I	10	530	5300	100	280900
J	15	542	8130	225	293764
	136	5454	75822	2320	2984092

$\bar{x} = 13.6 \quad \bar{y} = 545.4$

Therefore:

$$b = \frac{75822 - (10 \times 13.6 \times 545.4)}{2320 - (10 \times 13.6 \times 13.6)}$$

$$= 3.5$$

$$a = 545.4 - (3.5 \times 13.6)$$

$$= 497.8$$

Placing these values in the straight line formula (y = a + bx) gives us y = 497.8 + 3.5x and this can now be plotted on the scatter diagram (Fig. 5.4).

It is now possible to apply a test to determine how well the variables are correlated. The measure usually adopted for this purpose is the 'sample coefficient of determination' (r^2) which is given by:

$$r^2 = 1 - \frac{\Sigma(y_E - y_A)^2}{\Sigma(y_E - \bar{y}_E)^2}$$

where y_A is the actual value of the sample data and y_E is the estimated value using the regression line.

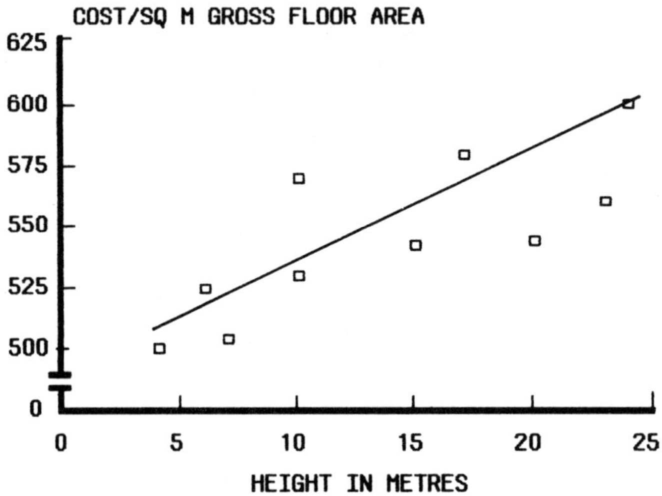

Figure 5.4 Scatter diagram (showing straight line of best fit)

Calculation of this expression can be quite tedious and short cut methods have been devised. One such method which uses the values of a and b already calculated is:

$$r^2 = \frac{a\Sigma y + b\Sigma xy - n\bar{y}^2}{\Sigma y^2 - n\bar{y}^2}$$

Using the previous data in the formula we get:

$$r^2 = \frac{(497.8 \times 5454) + (3.5 \times 75822) - (10 \times (545.4)^2)}{2984092 - (10 \times (545.4)^2)}$$

$$= 0.61$$

Thus about 60% of the total variation in the dependent

variable is explained by the relationships between y and x expressed in the regression line.

Another measure often used is the 'coefficient of correlation' which is the square root of the coefficient of determination. This is a measure of the probability that there is a genuine relationship between the variables. The nearer the value of the coefficient of correlation to plus or minus one, the narrower the range of the predicted values of the dependent variable.

The above example deals only with a two variable linear regression, but in practice it is quite usual to find that quite a large number of factors influence the cost of a building. The example assumed that the cost per square metre of floor varied only with the height so that it would be necessary to assume that other items such as shape, partitioning, standard of finish, level of services, etc., were constant. This would limit the application of the results to a very narrow range of buildings, but by extending the principle it is possible to increase the number of variables. In this case the technique is known as multiple linear regression. If we have a problem in which there are three variables it is still possible to draw a regression plot on a three dimensional scatter diagram and choose a plane of best fit rather than a line. Above three variables it is not possible to produce a graphical solution and because of the number of calculations involved a computer program is advisable to establish the coefficients (b) to be used in the regression formula which now becomes:

$$y = a + b_1 x_1 + b_2 x_2 + b_3 x_3 + \ldots + b_n x_n$$

where x is the value of each respective independent variable. The number of 'b' coefficients which have to be calculated will depend on the number of variables chosen.

<u>Description of program 5</u>

The program listed at the end of this chapter is for a three variable regression analysis.

Line 1015 Dimensions the three arrays which will contain the three variables.

Note that the array Y() contains the dependent variable and is referred to as Y1 in the prompts on the screen.

Lines 1025-1035 Initialise the variables to be used in the program (see Chapter 11).

Lines 1045-1050 Print the general heading on the printer.

Lines 1060-1155 Print the general instructions and permit the entry of data. Lines 1075-1085 together with the data in lines 10000 and 10005 are for testing purposes and can be removed when your program is working correctly.

Lines 1160-1165 Ask if a data check is required. If it is, the list of data is printed out on the printer using the subroutine at lines 1685-1745.

Lines 1175-1320 Deal with amendments to the list of data. If no amendments are required the program jumps to line 1330 at which there is a further option to obtain a data check.

Lines 1185-1225 list the editing options on the screen and allow selection.

Option 1. Alterations - uses lines 1235-1290.

Option 2. Additions - uses the original input line 1110.

Option 3. Deletion - uses lines 1300-1320.

Option 4. Finish editing.

Lines 1330-1335 Again ask if a data check is required. If the answer to the question is 'Y', the data list is printed out again on the printer but on this occasion there is no opportunity for further revisions.

Lines 1345-1385 Carry out the calculations. If perfect correlation between two variables is detected or if all the

	values of one variable are zero the program jumps to lines 1755-1780 which terminate the run after displaying appropriate text.
Lines 1395-1525	Print out the results of the regression on the printer.
Lines 1540-1545	Ask if estimates for Y1 are required.
Lines 1550-1615	If estimates for Y1 are required this section asks for values of X2 and X3, then calculates the estimated value of Y1 and prints the results on the printer. This final section of the program can be repeated as often as required to produce estimated values of Y1. When no further estimates are required, an opportunity is given to rerun the program.

Subroutines

Lines 1625-1645	Accept a line of data and allocate it to the correct elements in the three data arrays.
Lines 1655-1675	Remove a line of data.
Lines 1685-1745	Print the data check on the printer.
Lines 1785-1790	Check 'Y/N' answers.
Line 1795	Opens the printer channel.
Line 1800	Closes the printer channel.

The remaining subroutines are described in Chapter 11.

Table 5.2 Test example for program

Building	Cost/sq.m (gross floor area) (Y1)	Height (m) (X2)	Plan/shape index (X3)
A	500	4	1.1
B	544	20	1.5
C	580	17	1.7
D	570	10	1.7
E	504	7	1.2
F	560	23	1.6
G	600	24	2.0
H	524	6	1.3
I	530	10	1.3
J	542	15	1.4

Using these figures should produce the following:

```
KEY TO VARIABLES
Y1 IS COST/SQ.M.
X2 IS HEIGHT IN METRES
X3 IS PLAN/SHAPE INDEX

REGRESSION EQUATION IS
Y1= 373.813465 + .064893156 *X2 + 115.340532 *X3

S.D.OF REGRESSION ESTIMATE IS: 6.27262737

MEAN OF Y1 VALUES = 545.4
S.D.OF Y1 VALUES = 30.7902584

COEFFICIENTS OF CORRELATION:

MULTIPLE CORRELATION:  .98536713
BETWEEN Y1 & X2 IS:    .78019793
BETWEEN Y1 & X3 IS:    .985326641

PERCENTAGE OF VARIATION EXPLAINED BY:

MULTIPLE REGRESSION MODEL IS:  97.0948381
Y1 = A+B*X2           MODEL IS:  60.8708811
```

Y1 = A+B*X3 MODEL IS: 97.086859

Putting in values for X2 and X3 to obtain predicted values of Y1 produces the following:

FOR ESTIMATE X2 = 15 AND X3 = 1.5

ESTIMATED VALUE OF Y1 IS: 547.797661

95% OF ESTIMATE OF Y1 IS BETWEEN 535.503331 AND 560.09211

FOR ESTIMATE X2 = 5 AND X3 = 1

ESTIMATED VALUE OF Y1 IS: 489.478463

95% OF ESTIMATE OF Y1 IS BETWEEN 477.184114 AND 501.77281

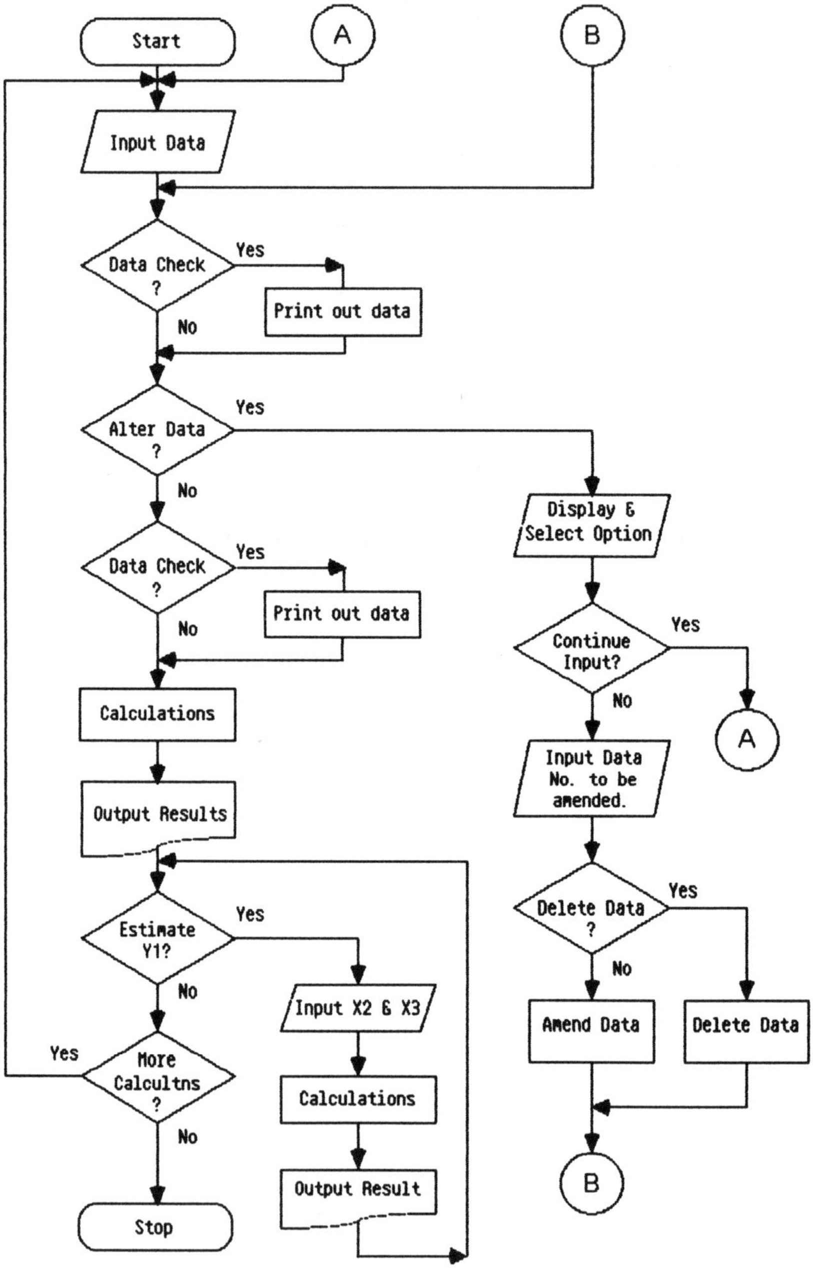

Figure 5.5 Flowchart for Program 5: Multiple regression analysis

```
1000 REM ***** MULTIPLE REGRESSION ANALYSIS PROGRAM.PROGRAM 5.*****

1005 REM ************ COPYRIGHT BRANDON, MOORE & MAIN *************
1010 REM *COLLINS PROFESSIONAL & TECHNICAL BOOKS (C)85. REV 6/2/85*
1015 DIM Y(50),X(50),Z(50):REM                      DIMENSION ARRAYS

1020 REM ******************** INITIALISE VARIABLES ****************
1025 SP$="":LT$="":FOR X= 1 TO 79:SP$=SP$+" ":LT$=LT$+"-":NEXT X
1030 CS$=CHR$(147):HS$=CHR$(19):RM$="E":DEL$=CHR$(20):CE$="@":LF$=""
1035 N=0:S1=0:S2=0:S3=0:Z1=0:Z2=0:Z3=0:Z4=0:Z5=0:T1=0:T2=0:T3=0

1040 REM *********************  PRINT HEADINGS  *******************
1045 GOSUB 1795:PRINT#2,"MULTIPLE REGRESSION PROGRAM"
1050 PRINT#2,LEFT$(LT$,27):PRINT#2,:GOSUB 1800

1055 REM ******************** INPUT DATA ROUTINE ******************
1060 GOSUB 4000:REM                                    CLEARS SCREEN
1065 PRINT"*** MULTIPLE REGRESSION PROGRAM ***":PRINT
1070 PRINT"OF TYPE:-    Y1 = A1 + B2*X2 + B3*X3":PRINT
1075 W$="COST/SQ.M":X$="HEIGHT  M":Y$="PLAN/SHAPE INDEX":REM****TEMP
1080 FOR L= 1 TO 10:READ Y1,X2,X3:REM                          ****TEMP
1085 GOSUB 1625:N=N+1:NEXT L:GOTO 1175:REM                     ****TEMP
1090 PRINT"AFTER EACH VARIABLE LISTED BELOW,"
1095 PRINT"TYPE-IN WHAT IT SYMBOLIZES.":PRINT
1100 PRINT"(EG: Y1 IS? 'THE COST/M2 OF HOUSING')":PRINT
1105 INPUT"Y1 IS";W$:INPUT"X2 IS";X$:INPUT"X3 IS";Y$
1110 RN=14:CL=0:GOSUB 5000:PRINT"INPUT VALUES OF Y1,X2,X3."
1115 PRINT"TO FINISH ENTER 'E'."
1120 PRINT"(NUMBER OF SETS OF DATA 3<N<50)":PRINT
1125 PRINT"NO.       Y1         X2         X3"
1130 RN=20:CL=0:GOSUB 5000:PRINT LEFT$(STR$(N+1)+")"+SP$,39)
1135 LL=9:I=1:D2=1
1140 CL=5:NO=0:GOSUB 6000:Y1=NX:IF Z$="E" THEN 1160
1145 CL=16:NO=0:GOSUB 6000:X2=NX:IF Z$="E" THEN 1160
1150 CL=27:NO=0:GOSUB 6000:X3=NX:IF Z$="E" THEN 1160
1155 L=N+1:GOSUB 1625:N=N+1:GOTO 1130:REM        ADD DATA TO TOTALS
1160 RN=22:CL=0:GOSUB 5000:PRINT"DO YOU WANT DATA CHECK? (Y/N)"
1165 GOSUB 1785:IF A$="Y" THEN GOSUB 1685

1170 REM *************** AMEND DATA OPTION MENU ******************
1175 RN=22:CL=0:GOSUB 5000:PRINT"ANY AMENDMENTS TO DATA? (Y/N)"
1180 GOSUB 1785:IF A$="N" THEN 1330
1185 GOSUB 4000:REM                                    CLEARS SCREEN
1190 PRINT"DO YOU WANT TO:"
1195 PRINT"     1. ALTER ANY DATA ITEMS?
1200 PRINT"     2. ADD   ..  ..  ..  .."
1205 PRINT"     3. DELETE..  ..  ..  .."
1210 PRINT"     4. FINISH EDITING DATA?":PRINT
```

```
1215 PRINT"ENTER SELECTION. (1-4)":PRINT
1220 GOSUB 5500:ZZ$=Z$:R=VAL(ZZ$):IF R<1 OR R>4 THEN 1220
1225 ON R GOTO 1235,1110,1300,1330


1230 REM ************** ALTER LINE OF DATA OPTION *****************
1235 PRINT"TYPE LINE NUMBER TO BE ALTERED":D2=0
1240 RN=9:CL=32:LL=2:I=0:NO=0:GOSUB 6000:L=NX:IF L<1 OR L>N THEN 1240
1245 GOSUB 1655:REM                              REMOVE DATA FROM TOTALS
1250 RN=11:CL=0:GOSUB 5000:PRINT"NOW INPUT CORRECT VALUES":PRINT
1255 PRINT"NO.       Y1          X2          X3"
1260 RN=15:CL=0:GOSUB 5000:PRINT LEFT$(STR$(L)+")"+SP$,39)
1265 LL=9:I=1:D2=0
1270 CL=5:NO=Y(L):GOSUB 6000:Y1=NX
1275 CL=16:NO=X(L):GOSUB 6000:X2=NX
1280 CL=27:NO=Z(L):GOSUB 6000:X3=NX
1285 GOSUB 1625:REM                                  ADD DATA TO TOTALS
1290 GOTO 1175


1295 REM ************** DELETE LINE OF DATA OPTION ****************
1300 PRINT"INPUT LINE NUMBER TO BE DELETED":D2=0
1305 RN=9:CL=32:LL=2:I=0:NO=0:GOSUB 6000:L=NX:IF L<1 OR L>N THEN 1305
1310 GOSUB 1655:REM                              REMOVE DATA FROM TOTALS
1315 FOR K= L TO (N-1):X(K)=X(K+1):Y(K)=Y(K+1):Z(K)=Z(K+1):NEXT K
1320 N=N-1:GOTO 1175


1325 REM **************** DATA CHECK ROUTINE **********************
1330 RN=22:CL=0:GOSUB 5000:PRINT"DO YOU WANT DATA CHECK? (Y/N)"
1335 GOSUB 1785:IF A$="Y" THEN GOSUB 1685


1340 REM *********************  CALCULATIONS ***********************
1345 C1=S1-(T1*T1/N):C2=S2-(T2*T2/N):C3=S3-(T3*T3/N)
1350 P3=Z3-(T1*T2/N):P4=Z4-(T1*T3/N):P5=Z5-(T2*T3/N)
1355 M1=T1/N:M2=T2/N:M3=T3/N:D=C2*C3-P5*P5
1360 IF D=0 THEN 1755
1365 B2=(P3*C3-P4*P5)/D:B3=(P4*C2-P3*P5)/D
1370 A1=M1-B2*M2-B3*M3:E=B2*P3+B3*P4:T=C1:R2=E/T
1375 R=SQR(R2):S=SQR((T*(1-R2))/(N-3)):D1=SQR(C1/N)
1380 R3=P3/SQR(C1*C2):R6=R3*R3
1385 R4=P4/SQR(C1*C3):R8=R4*R4


1390 REM ******************** PRINT RESULTS **********************
1395 GOSUB 1795
1400 PRINT#2,"KEY TO VARIABLES";Z$:PRINT#2,LEFT$(LT$,16);Z$
1405 PRINT#2,"Y1 IS ";W$;Z$:PRINT#2,Z$
1410 PRINT#2,"X2 IS ";X$;Z$:PRINT#2,Z$
1415 PRINT#2,"X3 IS ";Y$;Z$:PRINT#2,Z$:PRINT#2,Z$
1420 PRINT#2,"REGRESSION EQUATION IS:";Z$:PRINT#2,LEFT$(LT$,23);Z$
1425 S1$="+":IF B2<0 THEN S1$="-"
1430 S2$="+":IF B3<0 THEN S2$="-"
1435 PRINT#2,"Y1=";A1;S1$;ABS(B2);"*X2 ";S2$;ABS(B3);"*X3";Z$
```

```
1440 PRINT#2,Z$:PRINT#2,Z$
1445 PRINT#2,"S.D.OF REGRESSION ESTIMATE IS:";S;Z$
1450 PRINT#2,Z$:PRINT#2,Z$
1455 PRINT#2,"MEAN OF Y1 VALUES =";M1;Z$:PRINT#2,Z$
1460 PRINT#2,"S.D. OF Y1 VALUES =";D1;Z$
1465 PRINT#2,Z$:PRINT#2,Z$
1470 PRINT#2,"COEFFICIENTS OF CORRELATION:";Z$
1475 PRINT#2,LEFT$(LT$,28);Z$
1480 PRINT#2,"MULTIPLE CORRELATION:";R;Z$
1485 PRINT#2,"BETWEEN Y1 & X2 IS:  ";R3;Z$
1490 PRINT#2,"BETWEEN Y1 & X3 IS:  ";R4;Z$
1495 PRINT#2,Z$:PRINT#2,Z$
1500 PRINT#2,"PERCENTAGE OF VARIATION EXPLAINED BY:";Z$
1505 PRINT#2,LEFT$(LT$,37);Z$
1510 PRINT#2,"MULTIPLE REGRESSION MODEL IS:";R2*100;Z$
1515 PRINT#2,"Y1 = A+B*X2          MODEL IS:";R6*100;Z$
1520 PRINT#2,"Y1 = A+B*X3          MODEL IS:";R8*100;Z$
1525 PRINT#2,Z$:PRINT#2,Z$:GOSUB 1800


1530 REM ****************** Y ESTIMATE ROUTINE *******************
1535 GOSUB 4000:REM                                     CLEARS SCREEN
1540 PRINT"DO YOU WANT AN ESTIMATE FOR Y1? (Y/N)":PRINT
1545 GOSUB 1785:IF A$="N" THEN 1605
1550 PRINT"INPUT VALUES OF X2 AND X3":PRINT"     X2           X3"
1555 RN=5:LL=9:I=1:D2=0
1560 CL=1:NO=0:GOSUB 6000:X2=NX
1565 CL=14:NO=0:GOSUB 6000:X3=NX
1570 Y1=A1+B2*X2+B3*X3:L1=Y1-1.96*S:L2=Y1+1.96*S
1575 GOSUB 1795
1580 PRINT#2,"FOR ESTIMATE X2 =";X2" AND X3 =";X3;Z$:PRINT#2,Z$
1585 PRINT#2,"ESTIMATED VALUE OF Y1 IS:";Y1;Z$:PRINT#2,Z$
1590 PRINT#2,"95% ESTIMATE OF Y1 IS BETWEEN  ";L1;"  AND  ";L2;Z$
1595 PRINT#2,Z$:PRINT#2,Z$:GOSUB 1800
1600 GOTO 1535
1605 PRINT"DO YOU WANT ANOTHER RUN? (Y/N)"
1610 GOSUB 1785:IF A$="Y" THEN 1025
1615 GOTO 9998


1620 REM ********* SUBROUTINE TO ACCEPT SET OF DATA  *************
1625 T1=T1+Y1:T2=T2+X2:T3=T3+X3
1630 S1=S1+Y1*Y1:S2=S2+X2*X2:S3=S3+X3*X3
1635 Z3=Z3+Y1*X2:Z4=Z4+Y1*X3:Z5=Z5+X2*X3
1640 Y(L)=Y1:X(L)=X2:Z(L)=X3
1645 RETURN


1650 REM ********* SUBROUTINE TO REMOVE LINE OF DATA *************
1655 Y1=Y(L):X2=X(L):X3=Z(L)
1660 T1=T1-Y1:T2=T2-X2:T3=T3-X3
1665 S1=S1-Y1*Y1:S2=S2-X2*X2:S3=S3-X3*X3
1670 Z3=Z3-Y1*X2:Z4=Z4-Y1*X3:Z5=Z5-X2*X3
1675 RETURN
```

```
1680 REM ***************** PRINT DATA CHECK **********************
1685 GOSUB 1795:PRINT#2,"CHECK OF DATA ITEMS:";Z$
1690 PRINT#2,LEFT$(LT$,20);Z$
1695 PRINT#2,"LINE         Y1";LEFT$(SP$,17);"X2";LEFT$(SP$,17);"X3"
1700 PRINT#2,LT$;Z$
1705 FOR K= 1 TO N
1710 B=Y(K):GOSUB 8500:L$=N$:B=X(K):GOSUB 8500:M$=N$
1715 B=Z(K):GOSUB 8500
1720 PRINT#2,RIGHT$(SP$+STR$(K),2);"     ";L$;LEFT$(SP$,10);M$;
1725 PRINT#2,LEFT$(SP$,10);N$;Z$
1730 NEXT K
1735 PRINT#2,Z$:PRINT#2,"NUMBER OF ITEMS IS:";N;Z$
1740 PRINT#2,Z$:PRINT#2,Z$:GOSUB 1800
1745 RETURN


1750 REM ***** REJECT PERFECT CORRELATION BETWEEN 2 VARIABLES ****
1755 PRINT"DATA IS UNACCEPTABLE. PROGRAM WILL NOT"
1760 PRINT"TAKE DATA IN WHICH TWO OR MORE"
1765 PRINT"VARIABLES HAVE EXACT LIN.RELATIONSHIP."
1770 PRINT"IT WILL NOT ACCEPT DATA IN WHICH ALL"
1775 PRINT"OF THE VALUES OF ONE VARIABLE ARE ZERO.":PRINT
1780 GOTO 1605
1785 GOSUB 5500:A$=Z$:IF A$<>"Y" AND A$<>"N" THEN 1785
1790 RETURN
1795 OPEN2,4:Z$=LF$:RETURN:REM                 OPEN PRINTER CHANNEL!!!
1800 PRINT#2:CLOSE2:RETURN:REM                CLOSE PRINTER CHANNEL!!!
```

The following subroutines are required. (See Chapter 11)

4000 Clear screen routine.
4500 Home screen routine.
5000 Cursor position routine.
5500 Input character routine.
6000 Numeric input routine.

Important!
==========
Line 6035 to be changed as follows:

```
6035 IF Z$=RM$ AND D2=1 AND N>3 THEN 6100
```

8500 Format number to two decimal places.
9998 End program routine.

```
10000 DATA500,4,1.1,544,20,1.5,580,17,1.7,570,10,1.7,504,7,1.2
10005 DATA560,23,1.6,600,24,2,524,6,1.3,530,10,1.3,542,15,1.4
```

Chapter Six

Cost-in-Use

The increasing emphasis on building conservation and improved financial management has focussed the attention of clients and construction professionals on the whole life cycle costs of building. This has meant a renaissance in the consideration of cost-in-use techniques and many government departments now require an evaluation of this type - particularly when it comes to energy calculations.

Cost-in-use calculations attempt to bring all initial and future costs to a common unit, either 'present value' or 'annual equivalent' in order that a comparison between one alternative and another can be made.

At the heart of cost-in-use calculations is the concept of 'discounting' of future costs. In order to compare costs to be paid in the future with initial costs they need to be 'discounted' or brought back to their present value. By discounting we mean that the future value should be reduced to the sum that we would need to invest now for it to accrue at compound interest to the future cost. This is a very simple operation to undertake using valuation tables.

At the heart of all the basic valuation tables is the formula for compound interest.

$$A = (1 + i)^n$$

For further reference on this subject see 'Cost Planning of Buildings' (5th edition) Ferry, D.J. and Brandon, P.S., Granada 1984. Chapter 8.

Where A = amount one pound will accrue to in n time periods.

i = rate of interest per time interval divided by 100.

n = number of time periods (usually years).

Therefore if the rate of interest is 10% and we want to know the amount this will accrue to over five years then:

$$A = (1 + 0.10)^5$$
$$= 1.61$$

One pound will be worth £1.61 at the end of five years at 10% interest. The calculation is shown diagrammatically in Fig. 6.1.

Now this is a useful table for calculating the effect of inflation. If we thought prices would rise by 10% per annum over the next five years then we would know that something costing a pound now will have risen by 61% at the end of the period.

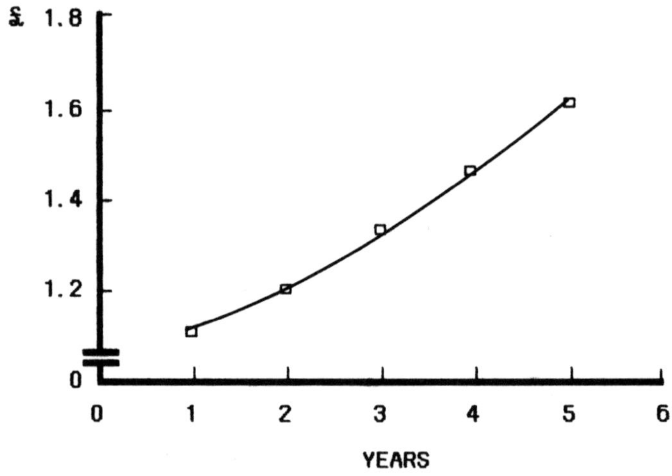

Figure 6.1 Amount of one pound at compound interest of 10% over five years

Discounting is merely the reverse of this process. The

formula for discounting a single amount at some time in the future is the reciprocal of the amount of one pound table. The formula is therefore:

$$\text{Present value (PV)} = \frac{1}{(1 + i)^n} \quad \text{or} \quad \frac{1}{A}$$

Where i = rate of interest divided by 100
n = number of time intervals.

Therefore if the rate of interest is 10% and we want to know the present value of one pound after five years then:

$$PV = \frac{1}{(1 + 0.10)^5}$$

$$= \frac{1}{1.61}$$

$$= 0.62$$

In other words the present value of one pound to be paid

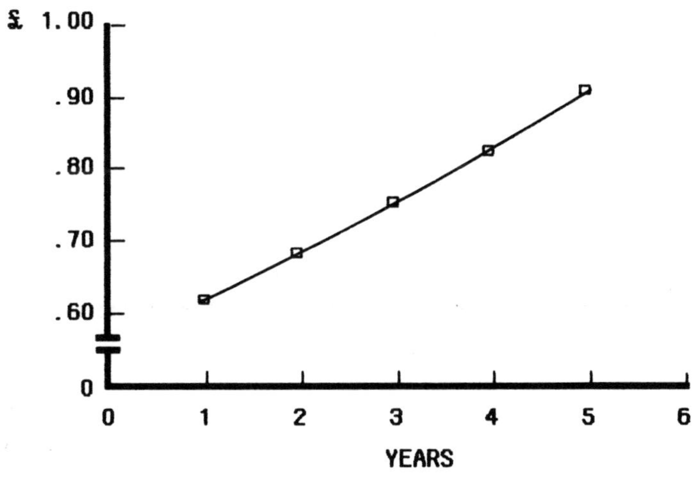

Figure 6.2 Present value of one pound over five years at a discount rate of 10%

or received in five years time at compound interest is sixty two pence. Figure 6.2 shows this calculation in diagrammatic form.

The principle behind all these tables is that all the expenditure on a building can be considered as an investment on which interest can be gained or lost. An amount to be paid or received in the future is worth less because a smaller sum invested now would accrue to the same amount at the future time period.

These two formulae can be brought together to allow for inflation and future cost. Suppose inflation is expected to rise at 5% per annum and the discount rate is 10%. Then the redecoration of a house costing £500 at present prices will have a value of £638.14 in five years time:

i.e. $£500 \times (1 + 0.05)^5 = 638.14$

and a present value of:

$£638.14 \times 1/(1 + 0.1)^5 = 395.65$

Shown diagrammatically the calculation looks like that shown in Fig. 6.3

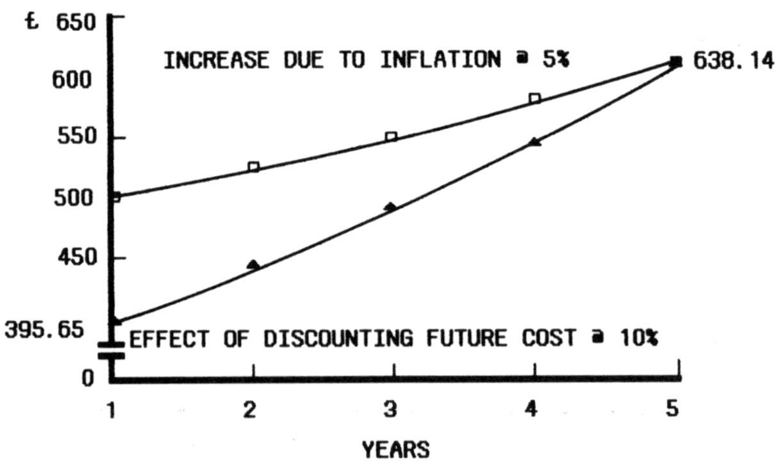

Figure 6.3 Effect of inflation rate of 5% and discount rate of 10% on a current value of £500 to be paid in five years' time

Practically all simple calculations in cost-in-use can be undertaken using just these two formulae although it is sometimes expedient to use other tables where the calculations would be long and tedious.

Where the present value is being calculated the other table of help to us is the present value of one pound per annum or 'years purchase' table. This table is described more fully in Chapter 7. What this table does is to find the present value of a regular series of payments in each time period in one operation. Whereas the previous present value table finds the value of a single payment or receipt sometime in the future, this new table finds the present value of the series.

There are a number of variations in the form of the years' purchase table but for our purposes in cost-in-use we shall concentrate solely on what is known as the single rate table. It is called this because the sinking fund included in the formula to replace the original capital has the same interest rate as the discount rate.

It should be borne in mind that although the tables can be very sophisticated there are an enormous number of assumptions made about the future in this type of calculation. The probability of error must be extremely high and therefore a rounding up or down to the nearest ten or one hundred pounds depending on the size of the problem should be considered to avoid a false impression of accuracy.

The formula for the present value of one pound per annum or years purchase is therefore:

$$\text{PV of £1 per annum} = \frac{(1 + i)^n - 1}{i(1 + i)^n}$$

Where i = rate of interest divided by 100

n = number of time intervals (usually years)

In normal practice, tables would be used to manually discover the value of the formulae for a given interest rate and number of time intervals. In computing it is no longer a problem to calculate the formula each time it is required and indeed it is an advantage where one or more of the variables are going to be changed during the evaluation process. It avoids having to look up the table to input the value or having it as a fixed value within the program.

An example of the use of the present value of one pound per annum table would be if we wanted to find the present value of the cleaning costs of a building over a forty year life span. If the cost were five thousand pounds per year then the total present value of the cleaning costs at 10% discount rate would be:

PV = £5000 x 9.78 where the value 9.78 has been calculated by the formula given above

= £48900

The total cleaning cost would be £200 000 but the high 10% discount rate reduces the future values considerably. This is an important fact to remember that the higher the discount rate the smaller will be the present value. Therefore a high interest rate will place more emphasis on initial costs rather than those that are incurred in the future.

The process of finding a total present value for a particular building or component is merely a question of adding together the initial cost with the sum of all the present values of future cost calculated by use of the present value and years purchase formulae to derive a single figure.

These techniques are the basis of the computer program described later. However one or two further techniques need to be understood to comprehend the listing.

Sensitivity analysis

One such technique is that of sensitivity analysis. The problem of predicting the future has already been mentioned. One method of assessing the effect of an error in one of the basic assumptions could be to repeat the calculation changing the value of the major variables by a predetermined amount each time. This is comparatively easy to undertake in a computer as a loop is built into the program and on each iteration of the loop a single variable is changed. Thus building life and interest rate could be changed and the resulting present values from each iteration could be plotted. Figure 6.4 shows such a plot for two components A and B for a change in these two variables.

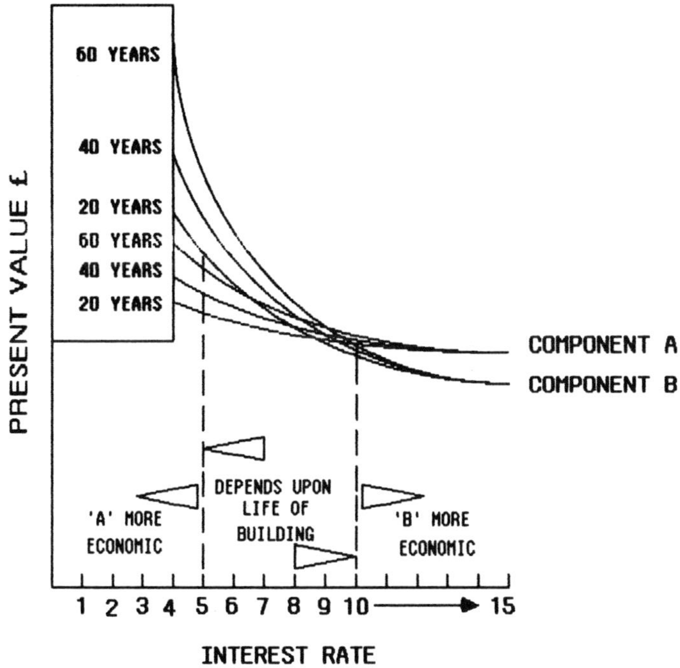

Figure 6.4 Sensitivity analysis

It is now possible to see how sensitive our results are to a marginal change in some of the basic assumptions - hence the name sensitivity analysis.

Taxation

Most commercial clients are in a position where they pay tax based on the profit of their organisation. The government in the UK and other industrial countries allows certain legitimate expenditure related to the business to be set against the profits made before tax is deducted. The actual allowable items will vary from country to country and from year to year depending on the legislation on tax which is passed at each budget. The following examples refer to the UK situation at the time of writing and it should be realised that it will be necessary to check the current figures and allowances before applying them to a real project.

Tax relief on initial cost of building

At the present time there are only a few types of building which attract tax relief on initial cost. These include research buildings and some agricultural buildings but the major sector to benefit is new industrial building. It is the industrial situation which we will use as our example but the same principles apply irrespective of the type.

On industrial buildings the government does not allow the whole of the initial cost to be set against profit in the first year (or until such time as sufficient profit is made for the client to benefit). Instead there is an initial allowance, known as initial writing down allowance, followed by a series of regular reliefs, known as further writing down allowances, until the whole of the building cost has been written off. At the present time the initial writing down allowance is 54% and further writing down allowances are at 4%. This leaves a balancing percentage of 2% in the twelfth year to make the full 100%.

The matter is complicated by the fact that there is delay between the time the expenditure is accounted for and the time the relief is actually received. This is because accounts are not prepared until the end of the financial year and it is usually several months before the accounts are agreed by the Inland Revenue. This delay is known as a tax lag and can be anything between one and three years in normal circumstances. Because this delays the benefit of the tax relief to the client it should be taken into account in our present value calculations.

The formula which will take account of the above factors will therefore be similar to the following (with reference to 'Building Design Evaluation. Cost in Use', by P.A. Stone, published by E. & F.N. Spon. Page 66).

$$\frac{T \times V^{TL}}{100}\left[\frac{IW}{100} + \frac{FW}{100}(V^{1} + V^{2} + \ldots + V^{n}) + \frac{BA \times V^{n+1}}{100}\right]$$

Where T = Tax rate
V = Present value
TL = Tax lag period in years
IW = Initial writing down allowance
FW = Future writing down allowances
BA = Balancing adjustment to write off 100% of building cost
n = Number of whole years to write off the remainder of the building cost

The reader should note that at the time of writing the government is rumoured to be contemplating the abolition of this tax relief.

Tax relief on maintenance and annual costs

This is a much simpler calculation as the total tax relief is allowable on the expenditure against the current year's profits. It is of course subject to the same tax lag as before and therefore the formula which needs to be applied to the discounted value of future cost is:

$$\frac{T}{100} \times V^{TL}(V^{1} + V^{2} + V^{3} + \ldots + V^{n})(\text{annual expenditure})$$

in other words all present values for future costs will be multiplied by $(T/100)xV^{TL}$

Where T = Tax rate
TL = Tax lag period in years
V = Present values for building life

Therefore all discounted values for a corporation tax rate of 52% and tax lag of one year will be reduced by

(52/100)(0.909)
= 0.473
assuming a discount rate of 10%.

This will apply to all maintenance and operational costs

and can be simply applied to the aggregated value of these costs once the present value calculation has been undertaken.

Grants

In addition to tax relief the government will also, in certain areas of the country and for certain industrial developments, provide a grant towards the initial cost of the building. This grant is usually provided soon after the expenditure is incurred and therefore there is not a similar time lag to that expected with tax relief.

The grant is therefore a straight deduction from the initial cost (before tax relief) and no further adjustment usually needs to be made.

Program 6A Cost-in-use demonstration model

Two programs have been written to illustrate the use of cost-in-use techniques applied to building incorporating the features described above. The first program is a simplified version of the second and is included to show more clearly the program techniques employed.

It relates to a single component but allows for taxation and sensitivity analysis. The operator has to provide a series of costings and other information related to the component as follows:

Note: A typical value is given in brackets for an industrial type building.

Corporation tax %	(52)
Initial writing down allowance %	(54)
Further writing down allowance %	(4)
Lag in tax (in years)	(1.5)
Rate of inflation %	(5)
Daily loss of production (The total cost of shutting the factory for a day)	(3000)
Grant %	(15)
Name of element/component	(Windows)
Initial cost	(100000)

Increase percentage (The extra cost of replacing as opposed to providing in the first place) %	(10)
Replacement period (The regular time interval between replacement of component)(in years)	(15)
Percentage replaced at each interval %	(10)
Disruption period (The time in days that the production will be affected by the replacement, etc.)	(20)
% loss of production (The degree to which the factory is disrupted by the replacement, etc. expressed as a percentage)	(10)
Annual regular maintenance costs (For example window cleaning)	(2000)
Regular maintenance periods (For example, painting every five years)	(5)
Regular maintenance costs (The cost of the painting every five years)	(10000)

In addition it is necessary to specify the range of values required for the sensitivity analysis. These relate to building life and the discount rate and it is necessary to give the minimum, maximum and step values required. For example for a building life you might choose 20, 60 and 20 as your input values which means that you will get three iterations on building life and for interest rate you might select 3, 15 and 3 as your inputs giving a further five iterations for each building life.

What the program does is to take all these values, calculate the cost-in-use as described previously in this chapter and then produce a table and a graph of the sensitivity analysis. It is possible by pressing 'RETURN' to alternate between the table and the graph.

A list of the output you should expect for the inputs

given above is shown in table 6.1 on page 106

Description of Program 6A

The following description of the program provides an outline of the programming method. It will also give guidance as to the techniques employed in the second cost-in-use program which may be more difficult to understand.

Lines 1000-1030	Dimension the arrays and set certain variable values. In particular line 1015 identifies the number of screen columns for the graph. Array P() stores the discounted present values for each year of the building life. Array R(,,) stores the results of each calculation. The first subscript defines whether or not the result includes for tax relief and grant, the second defines the discount rate and the third the building life. Array A$() is used to display a graph of the results.
Lines 1045-1150	Input values to all the main variables required in the calculations. For an example set see pages 100 & 101.
Lines 1160-1170	Print the headings for the final sensitivity table.
Lines 1180-1185	Mark the beginning of the iteration process for the sensitivity analysis.
Line 1190	Calculates the present value table for each year of the building life at the discount rate currently under consideration and places it in the array P. It allows for the inflation expected by combining inflation and present value tables together before

placing in P.

Line 1195 — Calculates the effect of the tax lag on the tax relief for the corporation tax.

Line 1205 — Calculates the replacement cost for one replacement including for disruption.

Line 1210 — Initialises some variables to be used in the following loop.

Lines 1215-1285 — This loop goes through each year of the building life and checks to see whether there is a replacement due or a regular maintenance period. It calculates the discounted value of these plus annual maintenance costs for that year, with or without tax relief.

Note: 1. There is no replacement or regular maintenance in the final year (line 1220).

2. There is no regular maintenance if more than 50% of the component is replaced (line 1245).

Lines 1295-1310 — Calculate tax relief on initial cost.

Line 1320 — Calculates the value of the grant and deducts from the total previously calculated.

Lines 1330-1350 — Print the results of each iteration as they are completed to give some display whilst the calculations are proceeding. Line 1330 rounds off the two values to be displayed to two decimal places. Variable 'TP' holds the present value and variable 'TT' holds the present value including tax relief and grant. Line 1350 proceeds with the next iteration.

Lines 1360-1375 — Find the highest calculated value so that the results may be scaled, the maximum value being placed in variable 'HI'. For the purposes of display, negative values are zeroed to give a vertical scale of zero to 'HI'.

Line 1380 — Variable 'HI' is increased by 5%, the interval being calculated as the increased value of 'HI' divided by 20. (A suitable value for the number of rows on most displays.) The number of columns available is then found by dividing the column width, minus ten characters for displaying the respective value, by the number of different discount rates used in the calculations.

Line 1385 — Initialises the array to be used to display the graph A$(), to blank spaces.

Lines 1390-1405 — Scale each result in turn and place an appropriate character at the correct position in the display array.

Firstly the vertical position is found, i.e. the number of the element in array A$() to which the result corresponds. This is calculated and stored in variable 'V'. The horizontal position is then calculated and stored in variable 'W'. This is the number of characters into the particular element of the array selected by the value of 'V'*. Finally a character is selected to be placed at the previously calculated position. This character corresponds to the set of results at a specified building life and is different depending on whether or not the set is inclusive of tax relief and grant.

* (Z-1) is used to minimise the number of overlapping results on the screen.

	The character is added into the result array at line 1400.
Lines 1410-1460	Display the graph of the results. This is done by displaying a value, initially the maximum value found (this being reduced by the value of the interval on successive iterations), followed by the corresponding element of the result array.
Lines 1465-1540	Re-display the table of results, this time each 'set' of results is given the associated character (see above). This display and the previous may now be alternated to allow the user to examine the results more closely. To exit from the program the user must type 'F' on the result display table.

Table 6.1 Output from the cost-in-use demonstration model Program 6A

```
COST     ! ELEMENT NAME: WINDOWS
 657604 !
 624724 !          C
 591844 !
 558964 !
 526084 !
 493204 !
 460324 !
 427444 !
 394564 !          B
 361684 !
 328804 !                      C
 295924 !
 263044 !         F            B
 230164 !
 197284 !          A                      C
 164404 !         E            A          A          C
 131524 !                     F                      A          C
  98644 !                     E          F
  65764 !         D           D          D          F          F
  32884 !
        !--------------------------------------------------------
INTEREST          3           6          9          12         15
PRESS THE RETURN KEY FOR TABLE.
```

BLDG. LIFE	DISC. RATE	PRESENT VALUE	INCLUDING TAX&GRANT
		GRP: A	GRP: D
20	3	208410	78658
20	6	178319	73249
20	9	158230	70019
20	12	144489	68169
20	15	134865	67195
		GRP: B	GRP: E
40	3	379307	158087
40	6	251978	109147
40	9	191477	86920
40	12	160154	76439
40	15	142544	71391
		GRP: C	GRP: F
60	3	626290	272880
60	6	313431	139095
60	9	207756	95196
60	12	164719	78850
60	15	143893	72127

PRESS RETURN FOR GRAPH, 'F' TO FINISH.

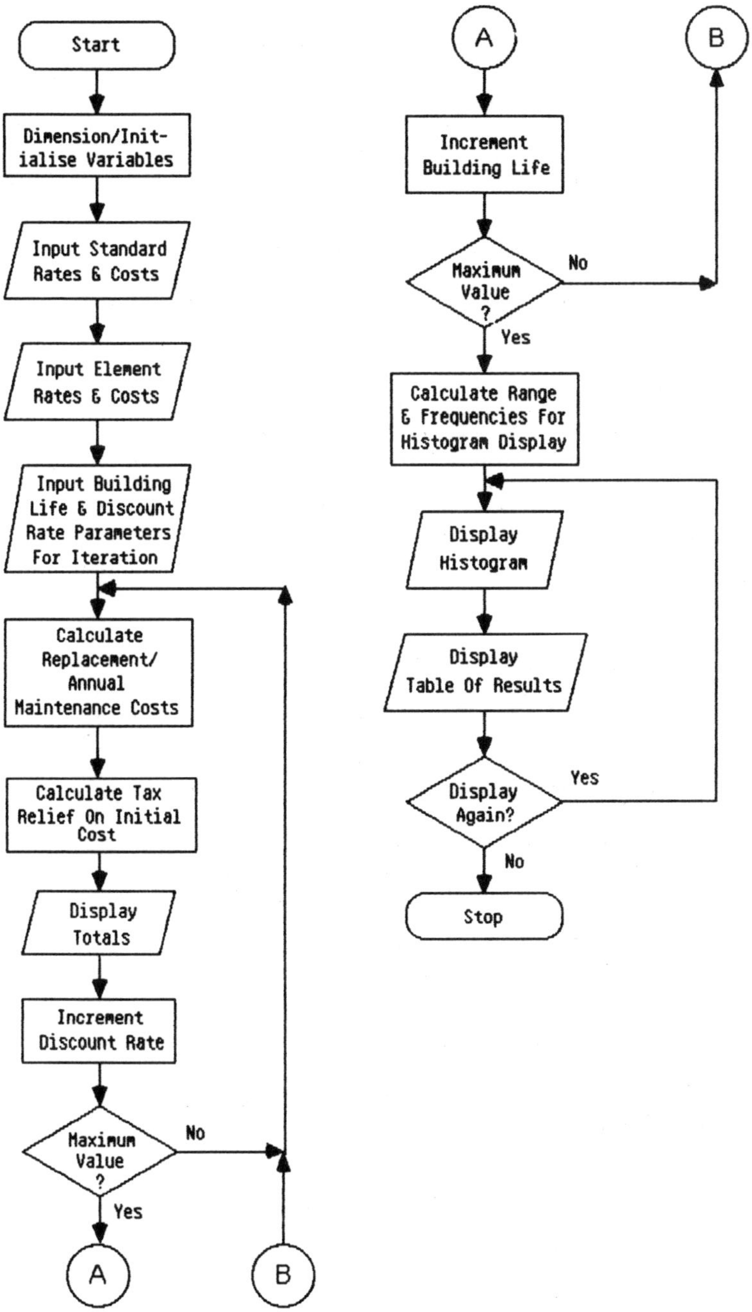

Figure 6.5: Flowchart for Program 6A: Cost-in-use demonstration model

```
1000 REM ****** COST-IN-USE DEMONSTRATION MODEL. PROGRAM 6A.******

1005 REM ************ COPYRIGHT BRANDON, MOORE AND MAIN **********
1010 REM *COLLINS PROFESSIONAL & TECHNICAL BOOKS (C)84 REV1/11/84*
1015 NL=80:REM***                          NO. OF COLUMNS ON MACHINE

1020 REM ************* DEFINE ARRAYS/VARIABLES ******************
1025 DIM P(100),R(2,10,10),A$(20)
1030 SP$="":LT$="":FOR X= 1 TO NL-10:SP$=SP$+" ":LT$=LT$+"-":NEXT

1035 REM ******************** INPUT DATA *************************
1040 PRINT CHR$(147);:REM                             CLEARS SCREEN!!!
1045 PRINT"COST-IN-USE EXAMPLE":PRINT
1050 INPUT"CORPORATION TAX                 (%)";CT
1055 INPUT"INITIAL WRITING DOWN ALLOWANCE(%)";FW
1060 INPUT"FURTHER WRITING DOWN ALLOWANCE(%)";DW
1065 INPUT"LAG IN TAX                    (YRS)";LG
1070 INPUT"RATE OF INFLATION               (%)";RI
1075 INPUT"DAILY LOSS IN PRODUCTION           ";OH
1080 INPUT"GRANT                           (%)";GT
1085 PRINT
1090 INPUT"ELEMENT NAME                       ";I$
1095 INPUT"INITIAL COST                       ";I1
1100 INPUT"INCREASE FOR REPLACEMENT        (%)";I2
1105 INPUT"REPLACEMENT PERIOD            (YRS)";I3
1110 INPUT"PERCENTAGE REPLACED             (%)";I4
1115 INPUT"DISRUPTION PERIOD            (DAYS)";I5
1120 INPUT"LOSS OF PRODUCTION              (%)";I6
1125 INPUT"ANNUAL REGULAR MAINTENANCE COST    ";I7
1130 INPUT"REGULAR MAINTENANCE PERIODS (YRS)";I8
1135 INPUT"REGULAR MAINTENANCE COSTS          ";I9
1140 PRINT
1145 INPUT"BUILDING LIFE  (MIN,MAX,STEP)      ";BM,BN,SB
1150 INPUT"DISCOUNT RATE  (MIN,MAX,STEP)      ";DM,DN,SD

1155 REM **************** PRINT TABLE HEADINGS ******************
1160 PRINT CHR$(147);:REM                             CLEARS SCREEN!!!
1165 PRINT"BLDG. DISC.  PRESENT  INCLUDING"
1170 PRINT"LIFE  RATE     VALUE  TAX&GRANT"

1175 REM *************** CALCULATE PRESENT VALUE ****************
1180 Q2=1:FOR BL=(BM) TO (BN) STEP (SB)
1185 Q1=0:FOR DC=(DM) TO (DN) STEP (SD)
1190 DF=(1+(RI/100))/(1+(DC/100)):FOR I= 1 TO BL:P(I)=DF↑I:NEXT I
1195 LT=DF↑LG:TF=LT*(CT/100)
```

```
1200 REM ******* CIU OF REPLACEMENT/ANNUAL MAINTENANCE COSTS******
1205 RS=I1*I2/100+I1:RQ=(RS*I4/100)+(OH*I5*I6/100):REM*REPLMT COST
1210 TP=I1:TT=0:K=INT(I3):J=INT(I8)
1215 FOR I= 1 TO BL
1220 IF I=BL THEN 1275:REM*** NO REPLACEMENT/REG. MAINT.IF LAST YR
1225 IF K<=0 THEN 1250
1230 IF I/K<>INT(I/K) THEN 1250
1235 PV=RQ*P(I):TP=TP+PV:REM                    *** CALC PRESENT VALUE
1240 TR=PV-(PV*TF):TT=TT+TR:REM                 *** CALC TAX RELIEF
1245 IF I4=>50 THEN 1275:REM*** NO REG. MAINT. IF 50%+ REPLACED.
1250 IF J<=0 THEN 1275
1255 IF I/J<>INT(I/J) THEN 1275
1260 RM=I9*P(I):TP=TP+RM:REM                    *** CALC REG. MAINT.
1265 TX=RM-(RM*TF):TT=TT+TX:REM                 *** CALC INC. TAX
1270 REM
1275 AM=I7*P(I):TP=TP+AM:REM                    *** CALC ANL. MAINT.
1280 TX=AM-(AM*TF):TT=TT+TX:REM                 *** CALC INC. TAX
1285 NEXT I


1290 REM ************* TAX RELIEF ON INITIAL COST ****************
1295 K=0:BJ=100-FW
1300 IF (BJ-DW)>=0 THEN BJ=BJ-DW:K=K+1:GOTO 1300
1305 YP=0:FOR I= 1 TO K:YP=P(I)+YP:NEXT I
1310 CW=I1*TF*(FW/100+((DW/100)*YP)+((BJ/100)*P(K+1))):TT=TT+I1-CW


1315 REM ******************* GRANT ALLOWANCE *********************
1320 TT=TT-(I1*GT/100)


1325 REM **************** PRINT TABLE OF RESULTS *****************
1330 TP=INT(.5+100*TP)/100:TT=INT(.5+100*TT)/100
1335 PRINT RIGHT$("   "+STR$(BL),3);RIGHT$("     "+STR$(DC),6);
1340 PRINT RIGHT$(SP$+STR$(INT(TP)),11);RIGHT$(SP$+STR$(INT(TT)),11)
1345 Q1=Q1+1:R(1,Q1,Q2)=TP:R(2,Q1,Q2)=TT
1350 NEXT DC:Q2=Q2+1:NEXT BL


1355 REM ********* CALCULATE RANGE/ VALUES FOR DISPLAY************
1360 HI=0:FOR X= 1 TO Q1:FOR Y= 1 TO Q2-1:FOR Z= 1 TO 2
1365 IF R(Z,X,Y)>HI THEN HI=R(Z,X,Y)
1370 IF R(Z,X,Y)<0 THEN R(Z,X,Y)=0
1375 NEXT Z:NEXT Y:NEXT X
1380 HI=INT(HI*1.05):RA=HI/20:NC=INT((NL-10)/Q1)
1385 FOR X= 1 TO 20:A$(X)=SP$:NEXT X
1390 FOR X= 1 TO Q1:FOR Y= 1 TO Q2-1:FOR Z= 1 TO 2
1395 V=INT(.5+R(Z,X,Y)/RA):W=NC*X-(Z-1):CH=(65-Q2+Y+Z*(Q2-1))
1400 A$(V)=LEFT$(A$(V),W-1)+CHR$(CH)+RIGHT$(A$(V),NL-10-W)
1405 NEXT Z:NEXT Y:NEXT X


1410 REM ****** DISPLAY GRAPHICAL INTERPRETATION OF RESULTS*******
1415 PRINT CHR$(147);:REM                              CLEARS SCREEN!!!
1420 PRINT"COST    ! ELEMENT NAME: ";I$
```

```
1425 VA=HI:FOR X= 20 TO 1 STEP-1
1430 PRINT RIGHT$(SP$+STR$(VA),7);" !";A$(X)
1435 VA=INT(.5+VA-RA)
1440 NEXT X
1445 PRINT"         !";LT$
1450 PRINT"INTEREST ";
1455 FOR X= DM TO DN STEP (SD):PRINT RIGHT$(SP$+STR$(X),NC);
1460 NEXT X:PRINT:PRINT"PRESS THE RETURN KEY FOR TABLE.";
1465 PRINT CHR$(19):REM                                 HOMES SCREEN!!!
1470 GET Z$:IF Z$<>CHR$(13) THEN 1470


1475 REM ******************* RE-DISPLAY TABLE *******************
1480 PRINT CHR$(147);:REM                               CLEARS SCREEN!!!
1485 PRINT"BLDG. DISC.  PRESENT  INCLUDING"
1490 PRINT"LIFE  RATE     VALUE  TAX&GRANT"
1495 X4=1:FOR BL=(BM) TO (BN) STEP (SB):X3=1
1500 PRINT"            GRP: ";CHR$(64+X4);"   GRP: ";CHR$(63+Q2+X4)
1505 FOR DC=(DM) TO (DN) STEP (SD)
1510 TP=R(1,X3,X4):TT=R(2,X3,X4)
1515 PRINT RIGHT$("   "+STR$(BL),3);RIGHT$("     "+STR$(DC),6);
1520 PRINT RIGHT$(SP$+STR$(INT(TP)),11);RIGHT$(SP$+STR$(INT(TT)),11)
1525 X3=X3+1:NEXT DC:X4=X4+1:NEXT BL
1530 PRINT:PRINT"PRESS RETURN FOR GRAPH, 'F' TO FINISH."
1535 GET Z$:IF Z$<>CHR$(13) AND Z$<>"F" THEN 1535
1540 IF Z$=CHR$(13) THEN 1415
1545 END
```

Program 6B Cost-in-use program

The second program is a more sophisticated version of the previous one allowing as it does for the evaluation of a complete industrial building. It also has a number of other interesting refinements which include the use of a disk operated data base in which previous evaluations can be stored, retrieved and amended.

In addition the operational costs of the building as a whole can be identified separately. These would include such costs as caretaking, insurances and possibly fuel cost although the latter could be dealt with under the element of heating if you so desired.

The program starts with a 'menu' which gives you the option of calling up a previous evaluation from disk or creating a new file (of project information). The new file consists of a table of input data and once this is complete it is possible to select from the menu the CALCULATION option which will then undertake the full cost-in-use evaluations. On completion this is filed away for future use and can then be called up and amended by typing over the previous figures as they appear on the screen.

The inputs are almost identical to those described for the demonstration model but are repeated here for clarity in the order and wording of this program.

For a new project the inputs are as follows:

(Note: INPUT abbreviations are used as the screen throughout)

1.	Name of file	A name for future retrieval should be included here
2.	Standard tax rate (%)	54
3.	Initial writing down allowance (%)	52
4.	Further writing down allowance (%)	4
5.	Lag in tax payment (years)	1.5
6.	Government grant (%)	15
7.	Inflation rate (%)	5
8.	Disruption cost (per day) due to loss in production	£1000
9.	Enter information relating to each element (A maximum of fifty elements can be used)	

An example of test data for five elements is shown in Table 6.2

10. Operational costs in /annum	Caretaking	£12000
(Note: up to 25 operational	Insurance	£3000
categories can be included)	Rates	£5000

Results

The results obtained from the cost-in-use program using the above input values should be as shown on page 123. These results are for the first iteration only.

Table 6.2 Showing input values for elements (test data only)

Element	Initial cost	Incr. (%)	Replacement period	% Replaced	Disruption period (days)	% Disrupted	Annual maint. cost	Period of regular maint. (years)	Periodical regular maint. cost
Foundation	15000	10	5	2	0	0	0	0	0
Frame	30000	0	0	0	0	0	0	0	0
Roof covering	10000	15	20	100	2	20	1000	5	500
Roof structure	18000	10	0	0	0	0	500	10	2000
Walls	40000	5	25	20	10	100	3000	5	8000

Description of Program 6B

The structure of the program is as follows but please refer to Chapter 11 for the more general input routines.

Lines 1020-1025 Dimension the arrays used in the program. These are:

F$() used to hold the index of files stored on disk. This array is stored along with the number of files, held in variable 'NF', in a directory file called 'CIUDIR'. As dimensioned the program will allow up to twenty files to be stored, this may be altered by changing the DIM statement and also the check at line 1160.

I$() stores the descriptions of the cost-in-use elements. Up to fifty elements may be used, but this number can be increased by changing all array assignments, in the DIM statements, that have a fifty included in them. Line 1420 will also have to be altered accordingly.

I(,) is a two dimensional array which holds the figures relating to each of the cost-in-use elements. As can be seen, each element has nine associated values, these are:

I(1,N) Initial cost.
I(2,N) Percentage increase in the initial cost with time (excluding 'normal' inflation).
I(3,N) Replacement period.
I(4,N) Percentage replaced.
I(5,N) Disruption period (days).
I(6,N) Disruption percentage.
I(7,N) Annual maintenance cost.
I(8,N) Regular maintenance period.
I(9,N) Regular maintenance cost.

P() this array stores the present

value table used in the calculations for each year of the building life, this being calculated at line 1710. At present the maximum building life is set to sixty years, this may be increased by merely altering this DIM statement.

TP() stores the present value totals for each element during the calculations at a certain building life and discount rate. These totals include for: increased costs, replacement costs, disruption costs due to replacement and regular and annual maintenance costs.

TT() this array is as TP() but all costs are inclusive of tax relief.

M$() stores the descriptions of the operational costs. The maximum number of operational elements is at present twenty-five but this number may be increased by altering in the DIM statement all assignments that have a twenty-five included in them. Line 1485 will also have to be altered accordingly.

M() holds the annual operational costs of the above elements.

TC() stores the present value totals for each of the operational elements.

TD() stores the present value totals for each of the operational elements including tax relief.

Lines 1035-1050 Initialise the standard variables used in the input routines (see chapter 11). Line 1050 defines a function to round off a number and then takes its integer.

Lines 1060-1080 Read from the disk the directory of files stored on disk placing them in array F$(). The number of files currently held is stored in variable 'NF'. The file 'CIUDIR' must be previously set up if using the program for the first time (see later).

Lines 1090-1170 Display the menu of options available to the user and wait until a selection is made by the input of a number between 1 and 6.

Lines 1150 and 1155 branch to the appropriate piece of coding for options 4 and 6.

Lines 1160 to 1170 check that if option 1 is chosen there are no more than twenty files already existing on the disk otherwise a message is displayed to the effect that no more files are allowed and the program is returned to the main menu.

Lines 1180-1220 Are the coding to allow the input of a file name (NA$). The index of existing files is then checked to see that:

(1) If option 1 is selected then the file name chosen does not clash with an existing one.

(2) If any of the other options are chosen, i.e. 2, 3 or 5, then the file does exist.

Once this check has been made the program branches to line 1525 in the case of option 3, line 1580 in the case of option 5 (after reading the specified file via the subroutine at line 2240) or continues on to line 1230.

Note: Users of some microcomputers, notably the BBC microcomputer, may find that it is necessary to initialise some variables to enable the program to run properly. The following extra line will cope with this problem.

1187 CT=0:FW=0:DW=0:LG=0:GT=0:JJ=0:KK=0:RI=0:OH=0

Lines 1230-1310 Input, or amend in the case of option 2, seven general parameters, these are:

Standard rate of tax	CT
Initial writing down allowance	FW
Further writing down allowance	DW
Lag in tax	LG
Initial grant	GT
Rate of inflation	RI
Loss in production (Disruption cost)	OH

Lines 1315-1445 Input or amend the cost-in-use data an element at a time. Variable 'J' is set to one at line 1310 and is incremented at line 1440 each time an element has been completed. This continues until 'J' becomes more than fifty (the maximum number of elements) or an '@' character is returned as an element description. When either of these conditions occurs the true number of elements is placed in variable JJ at line 1445.

Lines 1450-1510 Input or amend the operational cost data an element at a time. Variable 'K' is set to one at line 1450 and is incremented at line 1505 in a similar manner to variable 'J' in the previous paragraph. The only difference being that the maximum number of operational costs is twenty-five and the actual number of different operational costs is stored in variable 'KK' at line 1510.

Line 1515 This line stores the information that has just been input/amended. It first calls the subroutine at line

2310 which writes the cost-in-use data to the disk under the name given to it earlier (variable NA$). It then updates the directory file by calling the subroutine at line 2280 and finally returns back to the main menu by using line 9020 to clear the memory of data and effectively re-runs the program from the start. (See Chapter 11.)

Lines 1520-1540 Erase a data file from the disk. Line 1525 'squeezes' the file out of the directory array, the total file count being reduced by one. Line 1535 deletes the actual data file from the disk and it is left to line 1540 to write the amended directory file to disk and return back to the main menu via line 9020. (See Chapter 11.)

Lines 1550-1570 Display the directory of files stored on disk. Line 1565 may be changed if there are to be more than twenty files on a disk to stop the display scrolling.

Lines 1580-2220 Are the coding to perform the calculations and to print the results.

Lines 1590-1605 The first step in the calculations phase is to ask the operator for the range and step size of the two components on which the sensitivity analysis is to be performed. These are the building life and discount rate. The subroutine at line 2345 asks for the required input, this being the minimum, maximum and a step size, the information is returned in variables R1, R2 and R3 respectively. At present these have been preset with suggested values, if these are not applicable then the user may use his own by clearing each number and entering a new value.

Note: The program is set up to allow a maximum of five iterations of the discount rate. This is because the final summary table uses the full width of a standard eighty column printer. If more iterations are required and the user has a printer capable of printing more than eighty columns per line then the following alterations could be included.

1. Line 1620 DIM R(n,Q2,2) where 'n' is the new maximum number of iterations.
2. Line 2180 A2$ requires elongating.
3. Line 2195 add extra CHR$(75)s.
4. Line 2205 alter the FOR-TO-NEXT loop accordingly.

Line 1620	Calculates the number of iterations to be performed on the building life and dimensions the result array R(,,) dynamically. If the user's computer does not allow this facility then variable 'Q2' in the DIM statement should be replaced with an appropriate value.
Line 1625	Opens a channel to the printer.
Line 1630	If the printer to be used does not line-feed automatically then it may be forced to do so by ensuring that variable LF$ is set to a line-feed character (CHR$(10)) in line 1045.
Lines 1635-1695	Print some of the static values used in the calculations.
Line 1700	Initialises the variables used in the sensitivity analysis to their starting values.
Lines 1710-1715	See lines 1190-1195 of description of Program 6A.
Lines 1725-1750	Print the dynamic variables used in

	the calculation of the iteration, i.e. the values of building life and discount rate.
Lines 1760-1890	Calculate the cost of replacement, regular maintenance and annual maintenance for each of the CIU elements. (See description of Program 6A.) The calculations are exactly the same as those performed in lines 1205 to 1310 of program 6A, except that the array values of each element have been used and the totals over all the elements have been summed in the following variables: Total initial costs S1 Total present value of above S2 Total PV including tax relief S3 Total tax relief S4 (See lines 1760 and 1885)
Lines 1895-1965	Print the results of the previous calculations.
Lines 1975-1995	Calculate the annual operational costs for the building summing the totals in the following variables: Total annual operational cost O1 Total present value of above O2 Total PV including tax relief O3
Lines 2000-2050	Print the results of the previous calculations.
Lines 2055-2120	Prepare and print a summary of the results of the two sets of calculations. Line 2060 calculates the grant to be received on the initial costs.
Lines 2125-2135	Place the totals of the calculations in the result array R(,,) and then proceed with the next iteration. This is done by adding a step and checking to see if the maximum has been reached, if not the program

	continues with the next calculation by branching back to line 1710.
Lines 2140-2220	Print a sensitivity analysis table based on all the iterations that have been performed. At line 2220 the program closes the printer channel and returns back to the main menu via line 9020 which clears the memory of all data - effectively re-running the program. (See Chapter 11.)

Subroutines

Lines 2240-2270	Load a CIU file in from the disk.
Lines 2280-2305	Write the directory file 'CIUDIR' to disk.
Lines 2310-2335	Write a CIU file to disk.
Lines 2345-2375	Are the routine to input the parameters for the sensitivity analysis. These are minimum, maximum and a step size which are returned in variables R1, R2 and R3 respectively. Note that there is a check at line 2355 to ensure that the maximum value input is greater than the minimum.
Lines 2385-2420	Are two subroutines which draw boxes around the intended input. The first subroutine titles the box with the contents of strings Q1$ and Q2$.

Important

When using this program for the first time it may be necessary to set up the index file 'CIUDIR' before running the program otherwise a 'file not found' or similar error may be incurred. To avoid this problem a small program needs to be typed in and executed independently. This opens the file 'CIUDIR' for writing, writes a zero to it and closes the file. Some examples for different machines follow below:

Commodore

```
1 DOPEN#1,"CIUDIR",W:PRINT#1,0;CHR$(13);:DCLOSE#1
```

or for BASIC 2.0 users:

```
1 OPEN1,8,5,"0:CIUDIR,S,W":PRINT#1,0;CHR$(13);:CLOSE1
```

Apple

```
1 PRINT CHR$(4);"OPEN CIUDIR"
2 PRINT CHR$(4);"WRITE CIUDIR"
3 PRINT 0;CHR$(13);
4 PRINT CHR$(4);"CLOSE CIUDIR"
```

BBC

```
1 X=OPENOUT("CIUDIR")
2 PRINT#X,0
3 CLOSE#X
```

Microsoft BASIC

```
1 OPEN"O",1,"A:CIUDIR"
2 PRINT#1,0;CHR$(13);
3 CLOSE#1
```

Table 6.3 Results obtained from Program 6B using input values given in Table 6.2

Building life = 20 yrs
Discount rate = 3 %

Calculation of present values for replacement/maintenance costs

No.	Elem.name	Init.cost	P.V.total	P.V.tot(tax)
1	foundatn	15000	16204	6655
2	frame	30000	30000	12240
3	rf.cover.	10000	36450	15829
4	rf.struct	18000	32737	13890
5	walls	40000	143057	62098
	Totals	113000	258447	110712

Calculation of present values for operational costs

No.	Oper.name	Annl.cost	P.V.total	P.V.tot(tax)
1	caretaker	12000	295512	131265
2	insurance	3000	73878	32816
3	rates	5000	123130	54694
	Totals	20000	492520	218775

	Total excl. grant & tax	Grant on init.only	Tax relief	Net total
Initial costs!	113000	16950	66895	29155
Present value!	750967	0	421480	329487

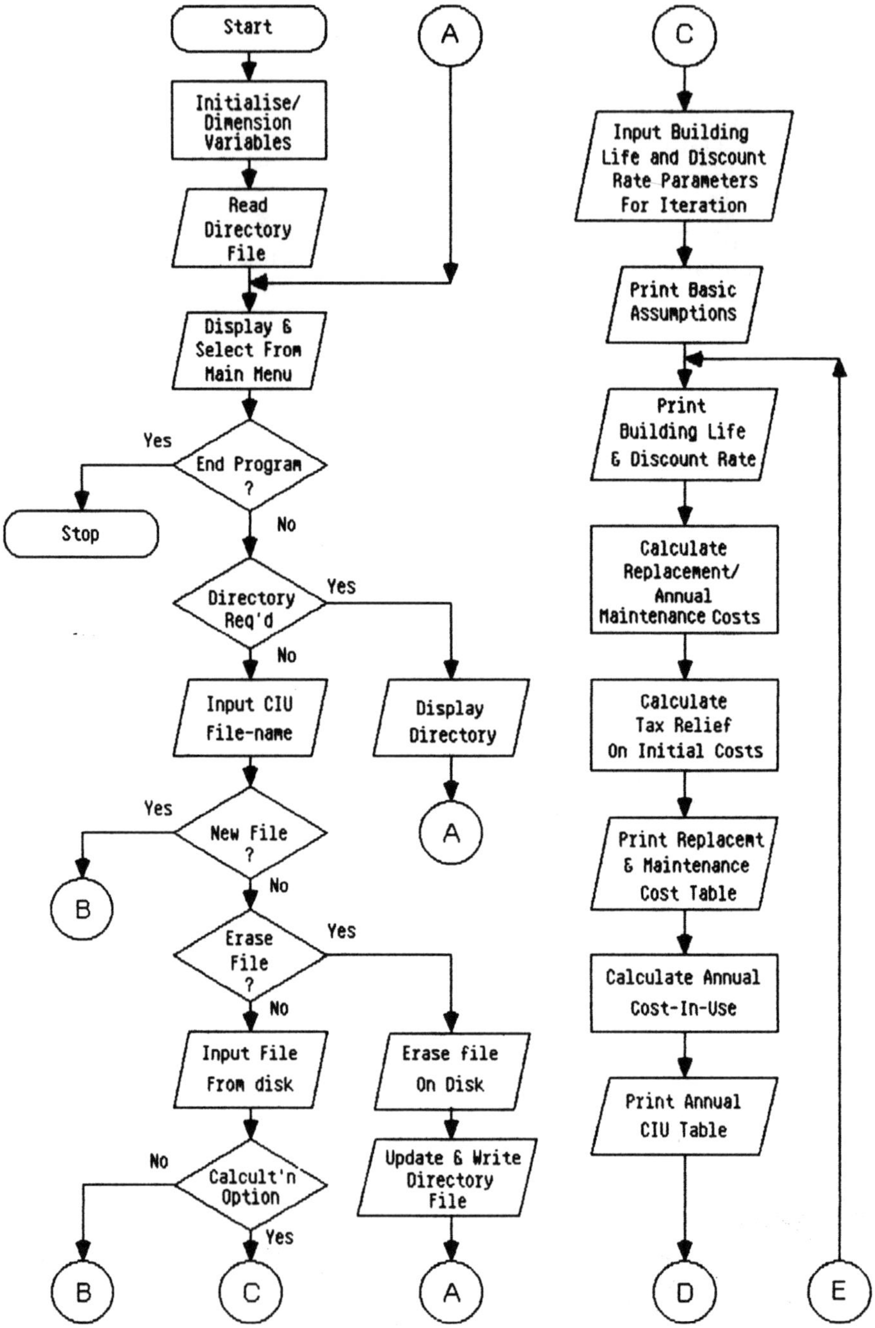

Figure 6.6 Flowchart for Program 6B: Cost-in-use

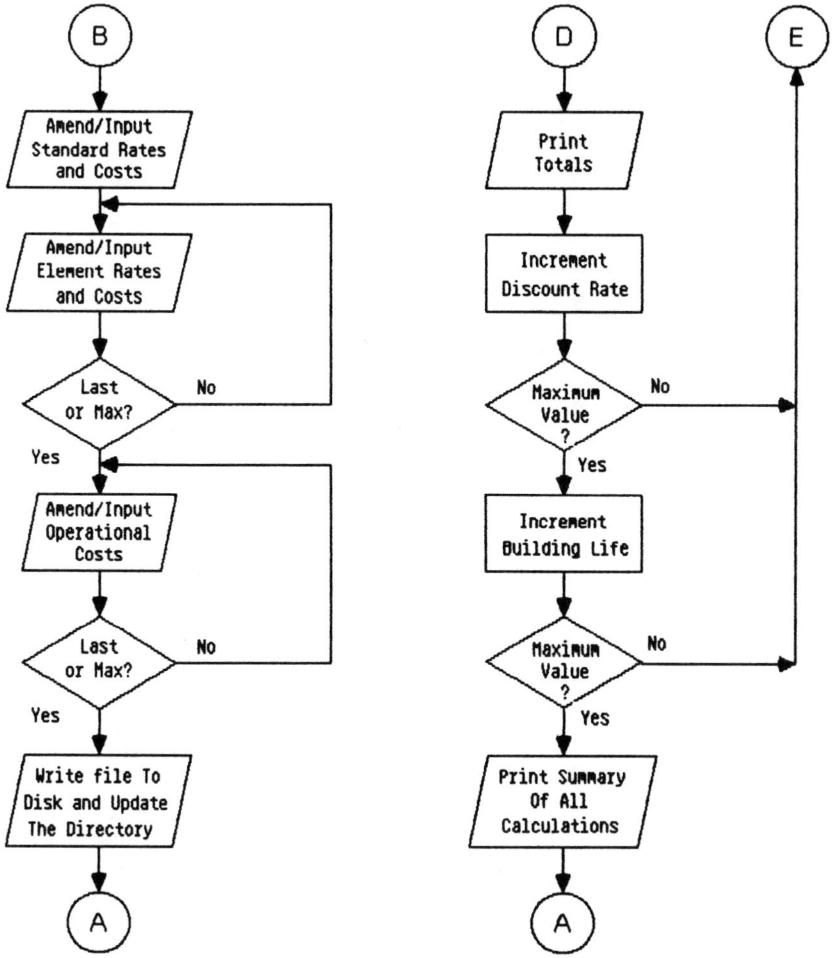

Figure 6.6 contd.

```
1000 REM *************** COST-IN-USE. PROGRAM 6B.*****************

1005 REM ************ COPYRIGHT BRANDON,MOORE AND MAIN ***********
1010 REM *COLLINS PROFESSIONAL & TECHNICAL BOOKS (C)84.REV9/10/84*

1015 REM **************** DEFINE ARRAYS TO BE USED  **************
1020 DIM F$(20),I$(50),I(9,50),P(60),TP(50),TT(50),M$(25),M(25)
1025 DIM TC(25),TD(25)

1030 REM **************** DEFINE MACHINE VARIABLES  **************
1035 LT$="":SP$="":FOR X= 1 TO 39:LT$=LT$+"-":SP$=SP$+" ":NEXT X
1040 CS$=CHR$(147):HS$=CHR$(19):RM$="↑":FC$="!":RV$="<":RO$=">"
1045 DEL$=CHR$(20):CE$="@":LF$="":RT$=CHR$(13):TD$="#":HL$=LT$
1050 DEF FNA(X)=INT(.5+X)

1055 REM **************** INPUT DISK DIRECTORY FILE **************
1060 DOPEN#1,"CIUDIR",D0:REM                            OPEN DISK!!!
1065 INPUT#1,NF:IF NF<>0 THEN 1075
1070 F$(1)="NO FILES ON DISK":GOTO 1080
1075 FOR X= 1 TO NF:INPUT#1,F$(X):NEXT X
1080 DCLOSE#1:REM                                      CLOSE DISK!!!

1085 REM *********************  MAIN MENU  ***********************
1090 GOSUB 4000:REM                                    CLEARS SCREEN
1095 PRINT"           ***";RV$;"COST-IN-USE PROGRAM";RO$;"***"
1100 PRINT:PRINT:PRINT RV$;"OPTIONS:";RO$
1105 PRINT LEFT$(SP$,11);RV$;"1";RO$;"-CREATE NEW FILE.":PRINT
1110 PRINT LEFT$(SP$,11);RV$;"2";RO$;"-AMEND OLD FILE.":PRINT
1115 PRINT LEFT$(SP$,11);RV$;"3";RO$;"-SCRATCH FILE.":PRINT
1120 PRINT LEFT$(SP$,11);RV$;"4";RO$;"-DIRECTORY OF DISK.":PRINT
1125 PRINT LEFT$(SP$,11);RV$;"5";RO$;"-CALCULATIONS.":PRINT
1130 PRINT LEFT$(SP$,11);RV$;"6";RO$;"-EXIT.":PRINT
1135 PRINT LEFT$(SP$,11);RV$;"SELECT FROM ABOVE";RO$
1140 GOSUB 5500:A$=Z$:A=VAL(A$):IF A=0 OR A>6 THEN 1140
1145 RN=17:CL=11:GOSUB 5000:PRINT RV$;"OPTION ";A$;" SELECTED";RO$
1150 IF A=6 THEN 9998
1155 IF A=4 THEN 1550
1160 IF (A=1 AND NF<20) OR A<>1 THEN 1180
1165 PRINT:PRINT RV$;"SORRY NO MORE FILES ALLOWED ON THIS DISK";RO$
1170 GOSUB 9000:GOTO 1090:REM                          PAUSE ROUTINE

1175 REM ******************  INPUT FILE-NAME  ********************
1180 QU$=" NAME OF FILE ":LL=12:C$="":RN=20:CL=0:GOSUB 9500
1185 GOSUB 7000:NA$=C$
1190 FOR X= 1 TO NF:IF NA$=F$(X) THEN 1205
1195 NEXT X:IF A=1 THEN NF=NF+1:F$(NF)=NA$:GOTO 1230
1200 ER$="NON-EXISTANT":QU=16:GOSUB 7505:GOTO 1180
```

```
1205 IF A=1 THEN ER$="ALREADY EXISTS":QU=16:GOSUB 7505:GOTO 1180
1210 IF A=3 THEN 1525
1215 GOSUB 2240:REM ***                    INPUT PROJECT FROM DISK
1220 IF A=5 THEN 1580


1225 REM ***************** INPUT / AMEND PROJECT *****************
1230 GOSUB 4000:REM                                  CLEARS SCREEN
1235 PRINT RV$;"CLIENT ATTRIBUTES";RO$:RN=5
1240 Q1$="STAND.TAX":Q2$="  RATE    ":LL=3:I=0:NO=CT:CL=0
1245 GOSUB 2385:GOSUB 6000:CT=NX
1250 Q1$="INIT.WRT.":Q2$="DWN.ALLOW":LL=3:I=0:NO=FW:CL=10
1255 GOSUB 2385:GOSUB 6000:FW=NX
1260 Q1$="FURT.WRT.":Q2$="DWN.ALLOW":LL=3:I=0:NO=DW:CL=20
1265 GOSUB 2385:GOSUB 6000:DW=NX:RN=10
1270 Q1$=" LAG IN   ":Q2$="TAX PAYMT":LL=5:I=1:NO=LG:CL=0
1275 GOSUB 2385:GOSUB 6000:LG=NX
1280 Q1$=" INITIAL ":Q2$=" GRANT(%)":LL=2:I=0:NO=GT:CL=10
1285 GOSUB 2385:GOSUB 6000:GT=NX:RN=15
1290 Q1$="RATE OF   ":Q2$="INFLATION":LL=3:I=0:NO=RI:CL=0
1295 GOSUB 2385:GOSUB 6000:RI=NX
1300 Q1$="LOSS IN   ":Q2$="PRODUCT'N":LL=6:I=0:NO=OH:CL=10
1305 GOSUB 2385:GOSUB 6000:OH=NX
1310 GOSUB 9000:J=1:REM                              PAUSE ROUTINE
1315 GOSUB 4000:REM                                  CLEARS SCREEN
1320 PRINT RV$;"INIT/REPLACEMENT ELEMENT COSTS";RO$
1325 PRINT:PRINT RV$;"ENTER '!' IN 1ST COL TO FINISH.";RO$
1330 Q1$=" ELEMENT:"+LEFT$(STR$(J)+" ",3):Q2$="    NAME     "
1335 C$=I$(J):LL=9:CL=1:RN=7
1340 GOSUB 2385:GOSUB 7000:IF C$="!" AND J>1 THEN 1445
1345 I$(J)=C$
1350 Q1$="INITIAL ":Q2$="COST     ":LL=6:I=0:NO=I(1,J):CL=20
1355 GOSUB 2385:CL=CL-1:GOSUB 6000:I(1,J)=NX
1360 Q1$="INCREASE":Q2$="   %     ":LL=3:I=0:NO=I(2,J):CL=30
1365 GOSUB 2385:GOSUB 6000:I(2,J)=NX:RN=14
1370 Q1$="REPLAMNT":Q2$="PERIOD   ":LL=2:I=0:NO=I(3,J):CL=1
1375 GOSUB 2385:GOSUB 6000:I(3,J)=NX
1380 Q1$="PERC'TGE":Q2$="REPLACED":LL=3:I=0:NO=I(4,J):CL=10
1385 GOSUB 2385:GOSUB 6000:I(4,J)=NX
1390 Q1$="DISRUPTN":Q2$="PERIOD   ":LL=2:I=0:NO=I(5,J):CL=20
1395 GOSUB 2385:GOSUB 6000:I(5,J)=NX
1400 Q1$="DISRUPTN":Q2$="PERC'TGE":LL=3:I=0:NO=I(6,J):CL=30
1405 GOSUB 2385:GOSUB 6000:I(6,J)=NX:RN=21
1410 Q1$="ANNUAL  ":Q2$="MAINTNCE":LL=5:I=0:NO=I(7,J):CL=1
1415 GOSUB 2385:GOSUB 6000:I(7,J)=NX
1420 Q1$="REG.MAIN":Q2$="PERIOD   ":LL=2:I=0:NO=I(8,J):CL=10
1425 GOSUB 2385:GOSUB 6000:I(8,J)=NX
1430 Q1$="REG.MAIN":Q2$="COST     ":LL=5:I=0:NO=I(9,J):CL=20
1435 GOSUB 2385:GOSUB 6000:I(9,J)=NX
1440 J=J+1:IF J<51 THEN 1315
1445 JJ=J-1
1450 K=1
1455 GOSUB 4000:REM                                  CLEARS SCREEN
1460 PRINT RV$;"ANNUAL OPERATIONAL COSTS/SALARIES";RO$:PRINT
1465 PRINT"ENTER '!' IN 1ST COL TO FINISH."
```

```
1470 PRINT"!---! !-----------! !--------!"
1475 PRINT"! NO! ! OPERATION ! ! ANNUAL !"
1480 PRINT"!---! !-----------! !--------!":RN=7:CL=0:GOSUB 5000
1485 PRINT"!";RIGHT$(" "+STR$(K),3);"!":PRINT"!---!"
1490 C$=M$(K):LL=9:RN=7:CL=6:QL=10:GOSUB 2410:GOSUB 7000:M$(K)=C$
1495 IF C$="!" AND K>1 THEN 1510
1500 NO=M(K):LL=6:CL=20:QL=7:I=0:GOSUB 2410:GOSUB 6000:M(K)=NX
1505 K=K+1:IF KK<26 THEN 1455
1510 KK=K-1
1515 GOSUB 2310:GOSUB 2280:Z$=RM$:GOTO 9020


1520 REM **************** SCRATCH PROJECT OPTION *****************
1525 FOR Y= X TO NF:F$(Y)=F$(Y+1):NEXT Y:NF=NF-1
1530 PRINT:PRINT LEFT$(SP$,10);RV$;"FILE BEING SCRATCHED";RO$
1535 SCRATCH(NA$),DO:REM                              SCRATCH FILE!!!
1540 GOSUB 2285:GOSUB 9000:Z$=RM$:GOTO 9020


1545 REM ****************** DIRECTORY OF DISK. *******************
1550 GOSUB 4000:REM                                    CLEARS SCREEN
1555 PRINT"           ***";RV$;"COST-IN-USE PROGRAM";RO$;"***":PRINT
1560 PRINT RV$;"DIRECTORY OF DATA DISK:";RO$
1565 FOR X= 1 TO NF:PRINT LEFT$(SP$,14);F$(X):NEXT X
1570 GOSUB 9000:Z$=RM$:GOTO 9020:REM                   PAUSE ROUTINE


1575 REM ****************** CALCULATION OPTION *******************
1580 GOSUB 4000:REM                                    CLEARS SCREEN
1585 PRINT"           ***";RV$;"CALCULATION OPTION";RO$;"***"
1590 PRINT:PRINT:PRINT RV$;"BUILDING LIFE";RO$
1595 RN=6:LL=3:R1=20:R2=60:R3=20:QU=16:GOSUB 2345:BM=R1:BN=R2:SB=R3
1600 PRINT:PRINT:PRINT RV$;"DISCOUNT RATE";RO$
1605 RN=11:LL=2:R1=3:R2=15:R3=3:QU=15:GOSUB 2345:DM=R1:DN=R2:SD=R3
1610 PRINT:PRINT:PRINT"PLEASE WAIT WHILE CALCULATIONS ARE MADE"
1615 PRINT:PRINT"TURN PRINTER ON..."
1620 Q2=INT(1+(BN-BM+1)/SB):DIM R(5,Q2,2)
1625 OPEN2,4:REM                               OPEN PRINTER CHANNEL!!!
1630 Z$=LF$
1635 PRINT#2,LEFT$(SP$,10);"COST-IN-USE PROGRAM";Z$
1640 PRINT#2,LEFT$(SP$,10);LEFT$(HL$,19);Z$:PRINT#2,Z$
1645 PRINT#2,"FILE-NAME: ";NA$;Z$:PRINT#2,LEFT$(HL$,10);Z$
1650 PRINT#2,"THE CALCULATIONS ARE BASED UPON ";
1655 PRINT#2,"THE FOLLOWING BASIC ASSUMPTIONS:-";Z$
1660 PRINT#2,HL$;LEFT$(HL$,26);Z$
1665 PRINT#2,"CORPORATION TAX.  ..  ..  .. = ";CT;"%";Z$
1670 PRINT#2,"INITIAL WRT'G DOWN ALLOWANCE = ";FW;"%";Z$
1675 PRINT#2,"FURTHER WRT'G DOWN ALLOWANCE = ";DW;"%";Z$
1680 PRINT#2,"LAG IN TAX PAYMENT..  ..  .. = ";LG;"YEARS";Z$
1685 PRINT#2,"RATE OF INFLATION ..  ..  .. = ";RI;"%";Z$
1690 PRINT#2,"LOSS IN PRODUCTION..  ..  .. = ";OH;"POUNDS";Z$
1695 PRINT#2,"GRANT ON INITIAL COSTS .  .. = ";GT;"%";Z$
1700 Q1=1:Q2=1:BL=BM:DC=DM
```

```
1705 REM *************** CALCULATE PRESENT VALUE *****************
1710 DF=(1+(RI/100))/(1+(DC/100)):FOR I= 1 TO BL:P(I)=DF↑I:NEXT I
1715 LT=DF↑LG:TF=LT*(CT/100)


1720 REM ********************** PAGE HEADINGS *******************
1725 PRINT#2,Z$:PRINT#2,Z$
1730 PRINT#2,"RESULTS ON THIS PAGE BASED UPON:-";Z$
1735 PRINT#2,LEFT$(HL$,33);Z$
1740 PRINT#2,"BUILDING LIFE = ";BL;"YRS";Z$
1745 PRINT#2,"DISCOUNT RATE = ";DC;"%";Z$
1750 PRINT#2,LEFT$(HL$,33);Z$


1755 REM ******* CIU OF REPLACEMENT/ANNUAL MAINTENANCE COSTS******
1760 S1=0:S2=0:S3=0:S4=0:FOR X= 1 TO JJ
1765 RS=I(1,X)*I(2,X)/100+I(1,X):REM         *** INCREASED COST
1770 RQ=(RS*I(4,X)/100)+(OH*I(5,X)*I(6,X)/100):REM* REPLACMNT COST
1775 TP=I(1,X):TT=0:K=INT(I(3,X)):J=INT(I(8,X))
1780 FOR I= 1 TO BL
1785 IF I=BL THEN 1840:REM*** NO REPLACEMENT/REG. MAINT.IF LAST YR
1790 IF K<=0 THEN 1815
1795 IF I/K<>INT(I/K) THEN 1815
1800 PV=RQ*P(I):TP=TP+PV:REM                 *** CALC PRESENT VALUE
1805 TR=PV-(PV*TF):TT=TT+TR:REM              *** CALC TAX RELIEF
1810 IF I(4,X)=>50 THEN 1840:REM*** NO REG. MAINT.IF 50%+ REPLACED
1815 IF J<=0 THEN 1840
1820 IF I/J<>INT(I/J) THEN 1840
1825 RM=I(9,X)*P(I):TP=TP+RM:REM             *** CALC REG. MAINT.
1830 TX=RM-(RM*TF):TT=TT+TX:REM              *** CALC INC. TAX
1835 REM
1840 AM=I(7,X)*P(I):TP=TP+AM:REM             *** CALC ANNUAL MAINT.
1845 TX=AM-(AM*TF):TT=TT+TX:REM              *** CALC INC. TAX
1850 NEXT I


1855 REM ************* TAX RELIEF ON INITIAL COST ****************
1860 K=0:BJ=100-FW
1865 IF (BJ-DW)>=0 THEN BJ=BJ-DW:K=K+1:GOTO 1865
1870 YP=0:FOR I= 1 TO K:YP=P(I)+YP:NEXT I
1875 CW=I(1,X)*TF*(FW/100+((DW/100)*YP)+((BJ/100)*P(K+1)))
1880 TT=TT+(I(1,X)-CW)
1885 TP(X)=TP:TT(X)=TT:S1=S1+I(1,X):S2=S2+TP:S3=S3+TT:S4=S4+CW
1890 NEXT X
1895 PRINT#2,Z$:PRINT#2,Z$
1900 PRINT#2,"CALCULATION OF PRESENT VALUES FOR REPLACEMENT/";
1905 PRINT#2,"MAINTENANCE COSTS";Z$:PRINT#2,HL$;LEFT$(HL$,24)
1910 A2$="!-----!-----------!-----------!--------------"
1915 A2$=A2$+"!--------------!"
1920 PRINT#2,"! NO. ! ELEM.NAME ! INIT.COST !   P.V.TOTAL   ";
1925 PRINT#2,"! P.V.TOT(TAX) !";Z$
1930 PRINT#2,A2$;Z$
1935 FO$=CHR$(69)+CHR$(138)+CHR$(75)+CHR$(14)+CHR$(14)
1940 FOR X= 1 TO JJ
1945 N(1)=X:N$(2)=I$(X):N(3)=I(1,X):N(4)=TP(X):N(5)=TT(X)
```

```
1950 GOSUB 8000:NEXT X:PRINT#2,A2$;Z$
1955 FO$=CHR$(144)+CHR$(75)+CHR$(14)+CHR$(14)
1960 N$(1)="TOTALS":N(2)=S1:N(3)=S2:N(4)=S3:GOSUB 8000
1965 PRINT#2,A2$;Z$


1970 REM ************* CALC.OF ANNUAL COSTS IN USE ***************
1975 YP=0:FOR I= 1 TO BL:YP=YP+P(I):NEXT I
1980 O1=0:O2=0:O3=0:FOR X= 1 TO KK
1985 TC(X)=M(X)*YP:O2=O2+TC(X):REM          *** OPERATIONAL COSTS
1990 TD(X)=TC(X)-(TC(X)*TF):O3=O3+TD(X):REM *** INCLUDING FOR TAX
1995 O1=O1+M(X):NEXT X
2000 PRINT#2,Z$:PRINT#2,Z$:PRINT#2,Z$
2005 PRINT#2,"CALCULATION OF PRESENT VALUES FOR OPERATIONAL COSTS"Z$
2010 PRINT#2,HL$;LEFT$(HL$,12);Z$
2015 PRINT#2,"! NO. ! OPER.NAME ! ANNL.COST !   P.V.TOTAL  ";
2020 PRINT#2,"! P.V.TOT(TAX) !";Z$:PRINT#2,A2$;Z$
2025 FO$=CHR$(69)+CHR$(138)+CHR$(75)+CHR$(14)+CHR$(14)
2030 FOR X= 1 TO KK
2035 N(1)=X:N$(2)=M$(X):N(3)=M(X):N(4)=TC(X):N(5)=TD(X):GOSUB 8000
2040 NEXT X:PRINT#2,A2$;Z$
2045 FO$=CHR$(144)+CHR$(75)+CHR$(14)+CHR$(14)
2050 N$(1)="TOTALS":N(2)=O1:N(3)=O2:N(4)=O3:GOSUB 8000


2055 REM ************** CALCS FOR SUMMARY SHEET *****************
2060 G1=S1*(GT/100):REM                     *** GRANT CALCULATION
2065 PRINT#2,A2$;Z$:PRINT#2,Z$:PRINT#2,Z$
2070 PRINT#2,"SUMMARY OF CALCULATIONS";Z$:PRINT#2,LEFT$(HL$,23);Z$
2075 FO$=CHR$(46)+CHR$(46)+CHR$(46)+CHR$(46):PRINT#2,Z$
2080 PRINT#2,LEFT$(SP$,16);"TOTAL EXCL.       GRANT ON       ";
2085 PRINT#2,"TAX RELIEF         NET TOTAL";Z$
2090 PRINT#2,LEFT$(SP$,16);"GRANT & TAX       INIT.ONLY";Z$
2095 PRINT#2,LEFT$(SP$,16);LEFT$(HL$+HL$,58);Z$
2100 PRINT#2,"INITIAL COSTS";:N(1)=S1:N(2)=G1:N(3)=S4:N(4)=S1-G1-S4
2105 GOSUB 8000:PRINT#2,Z$:PRINT#2,Z$
2110 PRINT#2,"PRESENT VALUE";:N(1)=S2+O2:N(2)=0:N(3)=S2+O2-S3-O3
2115 N(4)=S3+O3:GOSUB 8000:PRINT#2,Z$:PRINT#2,Z$
2120 PRINT#2,HL$;HL$;Z$:PRINT#2,Z$:PRINT#2,Z$
2125 R(Q1,Q2,1)=S2+O2:R(Q1,Q2,2)=S3+O3
2130 DC=DC+SD:IF DC<=DN THEN Q1=Q1+1:GOTO 1710
2135 BL=BL+SB:IF BL<=BN THEN Q2=Q2+1:Q1=1:DC=DM:GOTO 1710
2140 PRINT#2,Z$:PRINT#2,Z$:PRINT#2,Z$
2145 PRINT#2,"SUMMARY OF RESULTS-SENSITIVITY ANALYSIS";Z$
2150 PRINT#2,HL$;Z$:PRINT#2,Z$
2155 PRINT#2,"!YEAR !    CIU    !";LEFT$(SP$,28);"DISCOUNT RATE";Z$
2160 PRINT#2,"!      !DESCRIPTION!";
2165 FOR X= DM TO DN STEP SD
2170 PRINT#2,LEFT$(SP$,6-LEN(STR$(X)));X;"    !";:NEXT X
2175 A2$="!-----!-----------!-----------!-----------!-----------!"
2180 A2$=A2$+"-----------!-----------!"
2185 PRINT#2,Z$:PRINT#2,A2$;Z$
2190 T$(1)="STD. RATE":T$(2)="STD.+ TAX"
2195 FO$=CHR$(138)+CHR$(75)+CHR$(75)+CHR$(75)+CHR$(75)+CHR$(75)
2200 BL=BM:FOR X= 1 TO Q2:FOR Y= 1 TO 2
```

```
2205 N$(1)=T$(Y):FOR W= 1 TO 5:N(W+1)=R(W,X,Y):NEXT W
2210 PRINT#2,"!";RIGHT$(SP$+STR$(BL),4);" ";:GOSUB 8000
2215 NEXT Y:BL=BL+SB:PRINT#2,A2$;Z$:NEXT X
2220 CLOSE2:Z$=RM$:GOTO 9020:REM           CLOSE PRINTER CHANNEL!!!


2225 REM ***************** PROGRAM SUBROUTINES ******************


2230 REM ******************************************************


2235 REM ************** INPUT A PROJECT FROM DISK ****************
2240 DOPEN#1,(NA$),DO:REM                               OPEN DISK!!!
2245 PRINT:PRINT "      ";RV$;"FILE BEING LOADED. PLEASE WAIT";RO$
2250 INPUT#1,CT,FW,DW,LG,GT,JJ,KK,RI,OH
2255 FOR X= 1 TO JJ:INPUT#1,I$(X)
2260 FOR Y= 1 TO 9:INPUT#1,I(Y,X):NEXT Y:NEXT X
2265 FOR X= 1 TO KK:INPUT#1,M$(X),M(X):NEXT X
2270 DCLOSE#1:RETURN:REM                               CLOSE DISK!!!


2275 REM *************** WRITE A PROJECT TO DISK ****************
2280 PRINT:PRINT:PRINT"PLEASE WAIT-FILE BEING WRITTEN TO DISK"
2285 Z$=RT$
2290 SCRATCH"CIUDIR",DO:DOPEN#1,"CIUDIR",DO,W:REM       OPEN DISK!!!
2295 PRINT#1,NF;Z$;:IF NF=0 THEN 2305
2300 FOR X= 1 TO NF:PRINT#1,F$(X);Z$;:NEXT X
2305 DCLOSE#1:RETURN:REM                               CLOSE DISK!!!
2310 SCRATCH(NA$),DO:DOPEN#1,(NA$),DO,W:REM            OPEN DISK!!!
2315 PRINT#1,CT;Z$;FW;Z$;DW;Z$;LG;Z$;GT;Z$;JJ;Z$;KK;Z$;RI;Z$;OH;Z$;
2320 FOR X= 1 TO JJ:PRINT#1,I$(X);Z$;
2325 FOR Y= 1 TO 9:PRINT#1,I(Y,X);Z$;:NEXT Y:NEXT X
2330 FOR X= 1 TO KK:PRINT#1,M$(X);Z$;M(X);Z$;:NEXT X
2335 DCLOSE#1:RETURN:REM                               CLOSE DISK!!!


2340 REM *********** MAX/MIN PARAMETERS FOR ITERATION ************
2345 QU$="MIN=":I=0:CL=0:GOSUB 9500:NO=R1:GOSUB 6000:R1=NX
2350 QU$=QU$+"  "+STR$(R1)+" MAX=":CL=0:GOSUB 9500:NO=R2
2355 GOSUB 6000:IF NX<R1 THEN ER$="ERROR":GOSUB 7505:GOTO 2370
2360 R2=NX:QU$=QU$+"  "+STR$(R2)+" IN STEPS":CL=0:GOSUB 9500:NO=R3
2365 GOSUB 6000:R3=NX:RETURN
2370 RN=RN-1:CL=0:GOSUB 5000:RN=RN+1
2375 FOR X= 1 TO 3:PRINT SP$:NEXT X:GOTO 2345


2380 REM ******************* DRAW INPUT BOXES *******************
2385 RN=RN-3:GOSUB 5000:RN=RN+3
2390 PRINT Q1$:PRINT TAB(CL);Q2$:QL=LEN(Q1$)-1
2395 PRINT TAB(CL);"!";LEFT$(LT$,QL);"!"
2400 PRINT TAB(CL);"!";LEFT$(SP$,QL);"!"
2405 PRINT TAB(CL);"!";LEFT$(LT$,QL);"!":CL=CL+2:RETURN
2410 RN=RN-1:GOSUB 5000:RN=RN+1:PRINT "!";LEFT$(LT$,QL+1);"!"
2415 PRINT TAB(CL);"!";LEFT$(SP$,QL+1);"!"
```

```
2420 PRINT TAB(CL);"!";LEFT$(LT$,QL+1);"!":CL=CL+2:RETURN
```

The following subroutines are required. (See Chapter 11)

4000 Clear screen routine.
4500 Home screen routine.
5000 Cursor position routine.
5500 Input character routine.
6000 Numeric input routine.
7000 Alphanumeric input routine.

Important!
==========
The following lines are to be added:

```
6032 IF Z$=TD$ AND A=2 AND J<>JJ THEN J=JJ:GOTO 1450
7017 IF Z$=TD$ AND A=2 AND J<>JJ THEN J=JJ:GOTO 1450
```

7500 Report error routine.
8000 Format and print line routine.
9000 Pause routine.
9500 Draw input box routine.
9998 End program routine.

```
          ***<COST-IN-USE PROGRAM>***


<OPTIONS:>
            <1>-CREATE NEW FILE.

            <2>-AMEND OLD FILE.

            <3>-SCRATCH FILE.

            <4>-DIRECTORY OF DISK.

            <5>-CALCULATIONS.

            <6>-EXIT.

            <OPTION 5 SELECTED>

!---------------!-------------!
! NAME OF FILE  !     EXAMPLE<!
!---------------!-------------!
```

Screen display 6.1

```
          ***<CALCULATION OPTION>***


<BUILDING LIFE>
!------------------------------!----!
!MIN=   20 MAX=   60 IN STEPS ! 20 !
!------------------------------!----!

<DISCOUNT RATE>
!-----------------------------!---!
!MIN=   3 MAX=   15 IN STEPS ! 3 !
!-----------------------------!---!

PLEASE WAIT WHILE CALCULATIONS ARE MADE

TURN PRINTER ON...
```

Screen display 6.2

```
<CLIENT ATTRIBUTES>
STAND.TAX INIT.WRT. FURT.WRT.
  RATE    DWN.ALLOW DWN.ALLOW
!--------!!--------!!--------!
!  54    !!  52    !!   4    !
!--------!!--------!!--------!
 LAG IN    INITIAL
TAX PAYMT  GRANT(%)
!--------!!--------!
!  1.50  !! 15     !
!--------!!--------!
RATE OF   LOSS IN
INFLATION PRODUCT'N
!--------!!--------!
!   5    !!   1000 !
!--------!!--------!
    <PRESS 'SPACE BAR' TO CONTINUE>
```

Screen display 6.3

```
<INIT/REPLACEMENT ELEMENT COSTS>

<ENTER '!' IN 1ST COL TO FINISH.>
  ELEMENT: 3          INITIAL   INCREASE
     NAME             COST         %
 !-----------!        !-------! !-------!
 ! RF.COVER. !        ! 10000 ! !  15   !
 !-----------!        !-------! !-------!


 REPLAMNT PERC'TGE    DISRUPTN  DISRUPTN
 PERIOD   REPLACED    PERIOD    PERC'TGE
 !-------!!-------!   !-------! !-------!
 ! 20    !! 100   !   !  2    ! !  20   !
 !-------!!-------!   !-------! !-------!


 ANNUAL   REG.MAIN    REG.MAIN
 MAINTNCE PERIOD      COST
 !-------!!-------!   !-------!
 !  1000 !!  5    !   !   500<!
 !-------!!-------!   !-------!
```

Screen display 6.4

Chapter Seven
Discounted Cash Flow

This chapter introduces the concept of discounted cash flow but before this can be understood further tables need to be introduced. In chapter 6 we have illustrated the use of two of the common valuation tables - compound interest and present value of one pound. Valuation tables can be found in many publications including 'Cost Planning of Buildings' by Ferry and Brandon, Appendix D, and these often contain additional tables such as future value of one pound invested at regular intervals, present value of one pound payable at regular intervals and sinking fund.

The basic formula used in valuation is that for compound interest which enables us to calculate how much a sum of money invested will earn over a period of years. It is assumed that the interest which is earned is added to the capital so that each year there is more money invested to gain interest.

Now suppose that instead of a single lump sum being invested, the same amount is invested annually. One hundred pounds invested will gain interest over the first year which will be added to the capital as before, but then the second hundred pounds will be invested and so on. To evaluate the total amount at the end of some given period we need another formula which is:

For further reference on this subject see 'Cost Planning of Buildings' (5th edition) Ferry, D.J. and Brandon, P.S., Granada 1984. Chapter 4

$$A = \frac{(1 + i)^n - 1}{i}$$

Where A = amount one pound regularly invested at the beginning of each time period will accrue to in n time periods.

i = rate of interest per time interval divided by 100.

n = number of time periods (usually years).

This is known as the amount of one pound per annum table.

If the rate of interest is 10% and £100 is invested annually for five years:

$$A = \frac{(1 + 0.10)^5 - 1}{0.10} = 6.11$$

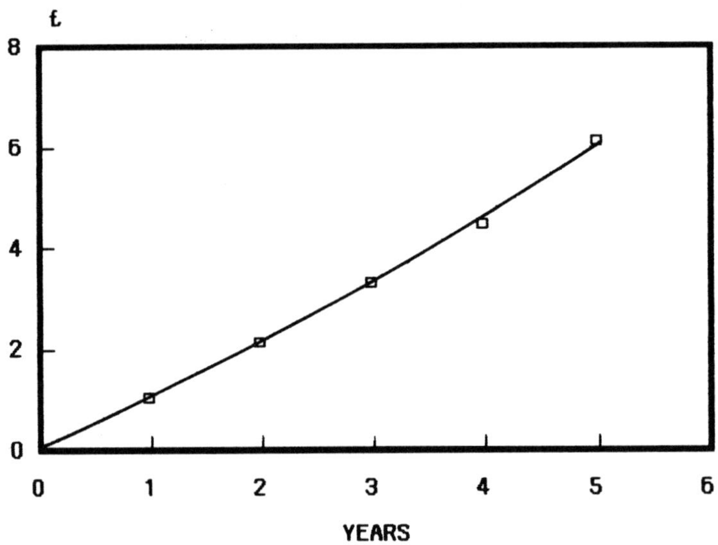

Figure 7.1 Future value of one pound invested at regular intervals at an interest rate of 10%

One pound invested each year for five years will be worth £6.11 so that £100 invested each year will be worth £611 assuming that the investment is made at the end of each year. If the investment is made earlier in the year it will be necessary to add on the interest gained by the first investment before the end of the first year. For example, suppose the investments are made six months before the end of each year it will be necessary to add:

$$\frac{£6.11 \times 10\%}{2} = £0.31$$

The total gained by one pound at the end of the fifth year will be £6.11 + £0.31 = £6.42

Sometimes we require the same information as above but the opposite way round, i.e. what sum of money must be invested annually in order to produce a given amount at the end of a period of time. This is necessary when building up a fund to replace a piece of equipment which is expected to be worn out in a given number of years. The formula for a 'sinking fund' is the reciprocal of the previous one:

$$S = \frac{i}{(1 + i)^n - 1}$$

Where S = amount of annual investment.
i = rate of interest per time interval divided by 100.
n = number of time intervals (usually years).

This is known as the annual sinking fund table.

Suppose we wish to accumulate £20000 at the end of fifteen years and the rate of interest is 10%:

$$S = \frac{0.10}{(1 + 0.10)^{15} - 1} = 0.031$$

So that 3.1 pence invested annually for fifteen years will be worth one pound at the end of the period hence the amount which must be invested each year will be:

£20000 x 0.031 = £620

Chapter 6 referred to the present value of one pound per annum or 'years purchase' table. This table is used to calculate the present value of future regular payments and the formula is obtained by combining the present value formula with that for the sinking fund to get:

$$PV = \frac{(1 + i)^n - 1}{i(1 + i)^n}$$

Where i = rate of interest per time interval divided by 100
n = number of time intervals (usually years)

If we pay the sum of £100 annually for ten years and the rate of interest is 10% the present value of one pound is:

$$\frac{(1 + 0.10)^{10} - 1}{0.10(1 + 0.10)^{10}} = 6.14$$

The present value of £100 paid annually over the period is therefore £614. This means that £614 invested now at 10% per annum would discharge the obligation of the ten annual payments of £100.

This chapter contains two programs, the first of these provides a substitute for valuation tables and uses the formulae given above and those included in Chapter 6. The second will calculate the internal rate of return on working capital.

Program 7A

This program will calculate:

1. Compound interest.
2. Future value of one pound invested at regular intervals.
3. Present value of one pound.
4. Present value of one pound payable at regular intervals.
5. Sinking fund.

The program is again menu driven and begins by displaying the above five options. The text for the screen display is put into the string array variables H$() and I$() at the beginning of the program in order that it can be used both to form the menu and as screen headings later in the program.

The GET statement is used to receive the menu choice, any value typed in other than the figures one to five is rejected. The numeric variable X is set to the chosen value, the screen is cleared (line 1095) and the appropriate elements of H$() and I$() are used to print the heading. In the case of option four, the additional text contained in lines 1110 and 1115 is printed.

All options require the number of periods and the percentage rate of interest per period to be known but after these data have been input the program splits to the relevant sections using the ON X statement in line 1140. The sections concerned are clearly labelled in the program - each section containing the required calculation in the line immediately following the REM statement. Following the calculation line is the section of program required to print out the results. As all the results need to be rounded to the nearest penny we have in this case used a defined function in line 1020 to do this. The function is named R and the numeric variable S in the brackets is set to whatever value is required when calling the function in the program; e.g. PRINT FNR(Q) causes the current value of Q to take the place of the numeric variable S in line 1020. The result of this calculation is then printed.

In the case of the first and second options the program asks for the amount invested and then multiplies this value by the result already obtained. All options end at line 1310 which allows the operator to recall the menu if desired.

If you use the examples given earlier in this book to test out this program you will find slight differences in the answers due to rounding. In the various examples given in the text above and in Chapter 6 the result obtained from each formula has been rounded before it has been multiplied by the sum involved. The computer program rounds the final answer only and is therefore mathematically more accurate.

Using the examples given in the text to test the various options in your computer program, the results should be as follows:

Option 1 Compound interest

Find the value of £1 invested at 10% compound interest after a period of five years (page 92).
Answer £1.61

If you enter the amount invested as £100 the total should come to £161.05

Option 2 Future value of £1 invested at regular intervals

Find the value of £1 invested annually for five years at 10% compound interest (page 136)
Answer £6.11

If you enter the amount invested as £100 the total should come to £610.51

Option 3 Present value of £1

If the rate of interest is 10%, find the present value of £1 after five years (page 93)
Answer 62 pence

Option 4 Present value of £1 payable at regular intervals. (Years purchase)

If the interest rate is 10%, find the present value of £1 paid annually for ten years (page 138)
Answer £6.14

If the sum paid annually is £100 the present value is £614.46

Option 5 Sinking fund

How much must be invested annually in order to accumulate £20000 at the end of fifteen years if the rate of interest is 10% (page 137)

To accumulate £1 at the end of the period the answer is 3.15 pence and for the total sum of £20000 it is £629.48.

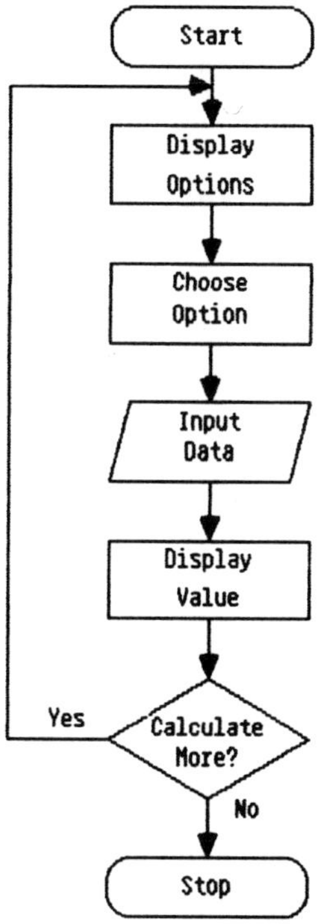

Figure 7.2 Flowchart for Program 7A: Discounting and cash flow

```
1000 REM ******** DISCOUNTING & CASH FLOW. PROGRAM 7A. ************


1005 REM *********** COPYRIGHT BRANDON, MOORE & MAIN **************
1010 REM *COLLINS PROFESSIONAL & TECHNICAL BOOKS (C)84  REV 7/9/84*
1015 DIM H$(5),I$(5):REM                                DIM ARRAYS
1020 DEF FNR(S)=INT(0.5+S*100)/100:REM                  ROUND RESULTS
1025 PRINT CHR$(147);:REM                               CLEARS SCREEN
1030 PRINT TAB(8);"DISCOUNTING & CASH FLOW":PRINT
1035 PRINT"THIS PROGRAM HAS THE FOLLOWING OPTIONS:":PRINT
1040 H$(1)="1) COMPOUND INTEREST.":I$(1)=""
1045 H$(2)="2) FUTURE VALUE OF 1 POUND INVESTED AT"
1050 I$(2)="   REGULAR INTERVALS."
1055 H$(3)="3) PRESENT VALUE OF 1 POUND":I$(3)=""
1060 H$(4)="4) PRESENT VALUE OF 1 POUND PAYABLE AT"
1065 I$(4)="   REGULAR INTERVALS.(YEARS PURCHASE)"
1070 H$(5)="5) SINKING FUND."


1075 REM ***************** DISPLAY MENU OF OPTIONS ****************
1080 FOR X= 1 TO 5:PRINT H$(X):PRINT I$(X):PRINT:NEXT X
1085 GET W$:X=VAL(W$):IF X<1 OR X>5 THEN 1085


1090 REM ************ INPUT TIME PERIODS/INTEREST RATE ************
1095 PRINT CHR$(147);:REM                               CLEARS SCREEN
1100 PRINT H$(X):PRINT I$(X):PRINT
1105 IF X<>4 THEN 1120
1110 PRINT"INCLUDING SINKING FUND AT THE"
1115 PRINT"SAME RATE AS INTEREST RATE":PRINT
1120 PRINT"THIS PROGRAM MAY BE USED IN RESPECT OF"
1125 PRINT"WEEKLY, MONTHLY OR YEARLY PERIODS.":PRINT
1130 INPUT"NUMBER OF PERIODS";N:PRINT
1135 INPUT"RATE OF INTEREST PER PERIOD";I:J=I/100:PRINT
1140 ON X GOTO 1150,1185,1210,1245,1260


1145 REM ***************** COMPOUND INTEREST *********************
1150 S=(1+J)↑N
1155 PRINT"VALUE OF 1 POUND AT THE END OF";N;"PERIODS"
1160 PRINT"AT";I;"% INTEREST=";FNR(S);" POUNDS":PRINT
1165 INPUT"AMOUNT INVESTED (POUNDS)";P
1170 PRINT P;"INVESTED HAS A VALUE OF:";FNR(P*S)
1175 GOTO 1310


1180 REM ******************** FUTURE VALUE ***********************
1185 S=((1+J)↑N-1)/J
1190 PRINT"VALUE OF 1 POUND AT END OF EACH OF";N;"PERIODS"
1195 PRINT"AT";I;"% INTEREST=";FNR(S);"POUNDS":PRINT
1200 GOTO 1165
```

```
1205 REM ******************** PRESENT VALUE ***********************
1210 S=(1/(1+J)↑N)
1215 INPUT"SUM (POUNDS)";P:PRINT
1220 PRINT"PRESENT VALUE OF ";P;"PAYABLE"
1225 PRINT"AT END OF";N;"PERIODS"
1230 PRINT"DISCOUNTED AT";I;"% = ";FNR(P*S);" POUNDS"
1235 GOTO 1310


1240 REM ************ PRESENT VALUE (YEARS PURCHASE) **************
1245 S=((1+J)↑N-1)/(J*(1+J)↑N)
1250 GOTO 1215


1255 REM ********************** SINKING FUND **********************
1260 S=J/((1+J)↑N-1):S=S*100
1265 PRINT"AMOUNT TO BE INVESTED AT END OF"
1270 PRINT N;"PERIODS TO ACCUMULATE 1 POUND BY END OF TERM"
1275 PRINT"AT";I;"% INTEREST=";FNR(S);"PENCE":PRINT
1280 INPUT"TOTAL TO ACCUMULATE (POUNDS)";T:PRINT
1285 Q=T*S/100
1290 PRINT"AMOUNT TO INVEST=";:IF FNR(Q)<1 THEN 1300
1295 PRINT FNR(Q);"POUNDS":GOTO 1310
1300 Q=Q*100:PRINT FNR(Q);"PENCE"


1305 REM **************** MORE CALCULATIONS ROUTINE ***************
1310 PRINT:PRINT"ANY MORE CALCULATIONS (Y OR N)?"
1315 GET W$:IF W$<>"Y" AND W$<>"N" THEN 1315
1320 IF W$="Y" THEN 1025
1330 END
```

Program 7B

When planning a project it is necessary to take into account not only the sums of money which will be involved but also the timing of the various payments and receipts, i.e. the cash flow. This concept of cash flow assumes that cash flow is positive when money is received and negative when money is paid out.

In order to calculate the 'internal rate of return' on working capital we need to try out various discount rates to find the rate which discounts the total receipts and the total expenditure to the same present day figure. The internal rate of return is the interest rate which when applied to the discounting of both positive and negative cash flows results in the same figure for each. Another way of expressing this is to say that the internal rate of return is the interest rate which produces a net present value (ie. positive minus negative discounted cash flows) equal to zero. To do this manually is very time consuming as it involves trial and error experiments with various discount rates.

There follows a simple program which can be used to try out various discount rates on a set of positive and negative cash flows. When all the data have been entered, the operator chooses a discount rate and the program produces the positive and negative cash flow totals together with the net present value.

```
10 DIM P(30),N(30)
20 PRINT CHR$(147);TAB(11);"CASH FLOW PROGRAM"
30 PRINT:PRINT:PRINT "HOW MANY TIME PERIODS";
40 INPUT T
45 T=T-1
50 PRINT CHR$(147);"   ENTER POSITIVE & NEGATIVE
   CASH FLOW"
60 PRINT:PRINT "   FOR EACH TIME PERIOD"
70 FOR K=0 TO T
80 PRINT:PRINT:PRINT " TIME PERIOD NO.";K
90 PRINT:PRINT:PRINT "   POSITIVE CASH FLOW";
100 INPUT P(K)
110 PRINT:PRINT:PRINT "   NEGATIVE CASH FLOW";
120 INPUT N(K)
```

```
130 PRINT CHR$(147)
140 NEXT K
150 PRINT CHR$(147):PRINT "WHICH DISCOUNT RATE";
160 INPUT D
165 I=D/100
170 PT=0:NT=0
180 FOR K=0 TO T
190 IF K=0 THEN 230
210 F=1/(1+I)↑K
220 GOTO 240
230 F=1
240 PT=(P(K)*F)+PT
250 NT=(N(K)*F)+NT
260 NEXT K
270 PRINT:PRINT:PRINT:PRINT "POSITIVE CASH FLOW TOTAL
    =";PT
280 PRINT:PRINT:PRINT:PRINT "NEGATIVE CASH FLOW TOTAL
    =";NT
290 PRINT:PRINT:PRINT:PRINT "NET PRESENT VALUE
    =";(PT-NT)
300 PRINT:PRINT:PRINT "ANOTHER DISCOUNT RATE (Y/N)";
310 INPUT A$
320 IF A$="Y" THEN 150
330 PRINT CHR$(147);:FOR I=1 TO 10:PRINT:NEXT I
340 PRINT "GOODBYE-PROGRAM FINISHED"
350 END
```

Program 7B is designed to automatically test a range of discount rates and hence calculate the internal rate of return on working capital. We have taken as an example a simple contract for a £410 000 building, to be constructed in thirty weeks. The prime cost to the contractor is £400 000 leaving £10000 profit. This represents 2.5% on turnover, but the real profit percentage may be very different from this.

In this example it is assumed that 10% of the value of the work is held as 'retention' by the client until the retention fund amounts to £25000 after which remaining work is paid for in full.

The same example is used in 'Cost Planning of Buildings' which has already been referred to earlier in this chapter. Table 7.1 gives the relevent data for this

Table 7.1 Data for test example

	Week no.	Actual cash flow Negative	Actual cash flow Positive
March	1	1000	
	2	1500	
	3	1500	
	4	2000	
April	5	3000	
	6	3000	
	7	3000	
	8	11000	8000
May	9	4000	
	10	4000	
	11	5000	
	12	14000	52000
	13	37000	
June	14	5000	
	15	6000	
	16	6000	
	17	67000	88000
July	18	5000	
	19	5000	
	20	4000	
	21	59000	70000
August	22	3000	
	23	3000	
	24	2000	
	25	29000	75000
	26	25000	
September	27	4000	
	28	4000	
	29	3000	
	30	41000	48000
October	34	33500	46500
May	64	5500	22500
	Total	400000	410000

example and is taken from Table 6 in the above book. Because the inputting of data takes quite some time, we have included as DATA in lines 10000 - 10040 all the information contained in Table 7.1. The temporary lines 1105 and 1110 READ the data and bypass the normal input routine. Once you have got your program typed in and running correctly, these lines can be removed and your own data typed in.

The routine used to enter data is similar to that used in Program 1 but this time three values are required for each reference number. These are the week number, negative cash flow and positive cash flow. Up to fifteen entries can be made per page after which there is an opportunity to amend any of the entries. When all data have been entered the next week number is entered as 'E' which terminates the entry routine and permits alterations to be made to the final page if this should be necessary.

As the calculations take some time a prompt appears on the screen stating that calculations are in progress. The weekly rate of interest is set in line 1030 with R0=0.1 being the lowest rate tested and R9=6 being the top rate to be tested.

The FOR-NEXT loop beginning in line 1205 is set to test in steps of 0.001% per week. The present value of cash flows is calculated, the negative values put into the numeric variable T1 and the positive into the numeric variable T2. The numeric array D() has two elements and is used to hold the difference between T1 and T2, the element D(2) holds the latest value and D(1) the previous value. The program uses the SGN(function in line 1220 to determine when one of the elements of D() is zero or when one is positive and the other is negative. Because D(1) will be zero at the beginning of the calculations, not yet having been given a value, this case is specifically excluded. When one of the elements is zero or a change in sign occurs we have obtained the required value of interest rate. A check is made to determine whether the current value of R or the previous value has given the answer nearest to zero. This value of interest rate is printed on the screen.

As the interest rate may be some way above the minimum, checks are made in lines 1245 and 1250. If the difference between the present value of the positive and negative cash flows is greater than 1000 the counter is raised by 0.1 instead of the normal 0.001. If the difference lies between 500 and 1000 the counter is raised by 0.05. This shortens the processing time considerably. If no result is found due to the interest rates being tested not including the required rate, the rates can be varied and the program rerun. Using the data given in the program gives an interest rate of 41.49% per annum. The difference between the present value of the positive and negative cash flows is 46 pence.

Line 1015	Dimensions array variables. C(,) will hold the input data. D() holds the difference between positive and negative cash flows as described above.
Lines 1020-1025	Initialise the general variables to be used in the program. (See Chapter 11.)
Line 1030	Sets the limits for the rates of interest to be tested.
Lines 1040-1100	Clear the screen and print the description of the program and operating instructions.
Lines 1105-1110	The temporary lines to read in the test data. These lines should be removed when the program is running satisfactorily.
Lines 1120-1180	These lines operate in conjunction with the subroutine at line 6000 for the inputting of data. Line 1160 keeps check of the cursor position and limits the number of entries per page to fifteen.
Lines 1190-1255	Carry out the calculations. Line 1205 sets up the loop to test the various interest rates. Lines 1210 and 1215 calculate the present value of positive and negative cash flows. Lines 1245 and 1250 artificially

	increase the value of the interest rate if the difference between positive and negative cash flows is greater than specified.
Lines 1265-1280	Print instructions for changing interest rates if no solution should be found.

Subroutines

Lines 1295-1310	Check Y/N answers.
Lines 1320-1345	Permit the alteration of data.

The remaining subroutines are described in Chapter 11.

Development

The program as written gives only the interest rate, the difference between the present value of the positive and negative cash flows is not printed. Try modifying the output to print this value for a revised discount rate and also the totals of the positive and negative cash flows.

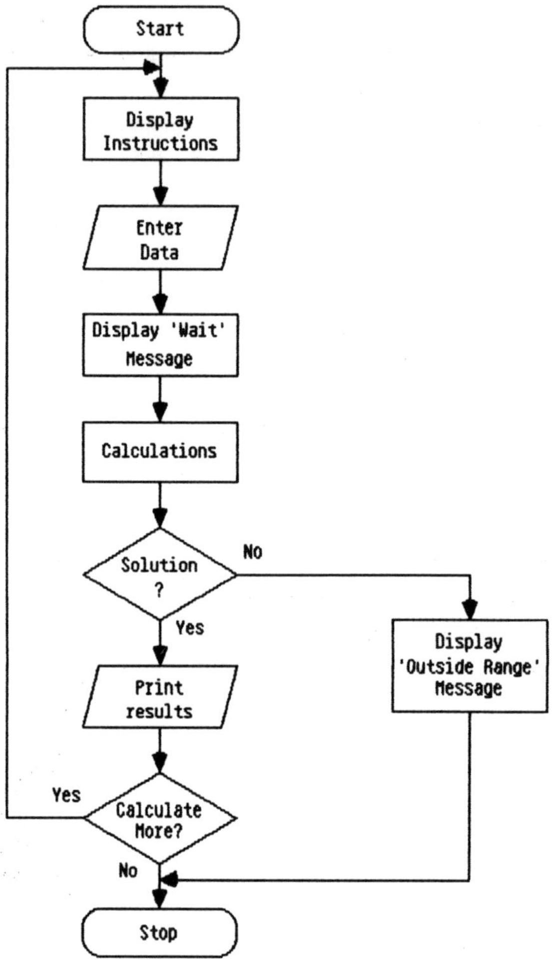

Figure 7.3 Flowchart for Program 7B: Rate of return

```
1000 REM ************** RATE OF RETURN. PROGRAM 7B. ***************

1005 REM ************ COPYRIGHT BRANDON, MOORE & MAIN *************
1010 REM *COLLINS PROFESSIONAL & TECHNICAL BOOKS (C)84 REV 23/5/84*
1015 DIM C(50,3),D(2):REM                           DIMENSION ARRAYS
1020 SP$="":FOR X= 1 TO 39:SP$=SP$+" ":NEXT X
1025 CS$=CHR$(147):HS$=CHR$(19):RM$="E":DEL$=CHR$(20):CE$="@"
1030 R0=.1:R9=6

1035 REM ************ DISPLAY EXPLANATION OF PROGRAM **************
1040 GOSUB 4000:REM                                    CLEARS SCREEN
1045 PRINT LEFT$(SP$,12);"RATE OF RETURN":PRINT
1050 PRINT"THIS PROGRAM IS USED TO CALCULATE THE"
1055 PRINT"RATE OF RETURN ON WORKING CAPITAL.":PRINT
1060 PRINT"ENTER THE WEEK BY WEEK ACTUAL"
1065 PRINT"CASH FLOW (NEGATIVE OR POSITIVE).":PRINT
1070 PRINT"THE MACHINE WILL CALCULATE THE"
1075 PRINT"PERCENTAGE RATE OF RETURN ON THE"
1080 PRINT"CONTRACTORS WORKING CAPITAL."
1085 PRINT"(RANGE";R0*52;"-";R9*52;"PER ANNUM)":PRINT
1090 PRINT"ENTER WEEK NUMBER AS 'E' TO FINISH.":PRINT:PRINT
1095 PRINT"TOUCH RETURN TO CONTINUE."
1100 GOSUB 5500:IF Z$<>CHR$(13) THEN 1100
1105 N=32:FOR J= 1 TO 32:FOR K= 1 TO 3:READ C(J,K):REM ******* TEMP
1110 NEXT K:NEXT J:REM                                  ******* TEMP

1115 REM ****************** ENTER DATA ROUTINE *******************
1120 GOSUB 4000:REM                                    CLEARS SCREEN
1125 PRINT"REF   WEEK NO.   NEGATIVE    POSITIVE":PRINT
1130 J1=1:N=0:RN=4:FOR J= 1 TO 50
1135 CL=0:GOSUB 5000:PRINT STR$(J);")":D2=1
1140 FOR K= 1 TO 3
1145 CL=8:I=0:LL=2:IF K>1 THEN I=1:LL=9:CL=K*12-8
1150 N0=C(J,K):GOSUB 6000:IF Z$="E" THEN J=50:GOTO 1175
1155 C(J,K)=NX:NEXT K:N=N+1:RN=RN+1
1160 IF RN<19 THEN 1175
1165 GOSUB 1295:J1=J1+15:REM                       AMENDMENT ROUTINE
1170 RN=4:CL=0:GOSUB 5000:FOR X= 1 TO 20:PRINT SP$:NEXT X
1175 NEXT J
1180 IF Z$="E" OR N=50 THEN GOSUB 1295:REM         AMENDMENT ROUTINE

1185 REM ********************* CALCULATIONS **********************
1190 GOSUB 4000:REM                                    CLEARS SCREEN
1195 RN=11:CL=0:GOSUB 5000
1200 PRINT"PLEASE WAIT - CALCULATIONS IN PROGRESS"
1205 FOR R= R0 TO R9 STEP .001:T1=0:T2=0
1210 FOR J= 1 TO N:X=1/(1+R/100)↑C(J,1)
1215 T1=T1+X*C(J,2):T2=T2+X*C(J,3):NEXT J
1220 D(2)=T1-T2:IF SGN(D(2))=SGN(D(1)) OR R=R0 THEN 1245
```

```
1225 IF ABS(D(1))>ABS(D(2)) THEN R=R-.01
1230 RN=11:CL=0:GOSUB 5000:PRINT SP$:GOSUB 5000
1235 PRINT"RATE OF RETURN =";INT(R*5200)/100;"% PER ANNUM"
1240 GOTO 9998
1245 D(1)=D(2):D(2)=0:IF ABS(D(1))>1000 THEN R=R+.1
1250 IF ABS(D(1))<1000 AND ABS(D(1))>500 THEN R=R+.05
1255 NEXT R
1260 GOSUB 4000:REM                                CLEARS SCREEN
1265 PRINT"NO SOLUTION - OUTSIDE RANGE OF INTEREST"
1270 PRINT"RATES. TO CHANGE RATES ALTER VALUES OF"
1275 PRINT"R0 & R9 IN LINE 125 THESE ARE WEEKLY"
1280 PRINT"RATES OF INTEREST."
1285 GOTO 9998


1290 REM ******************* CHECK SUBROUTINE *********************
1295 RN=22:CL=0
1300 GOSUB 5000:PRINT"ANY ALTERATIONS?(Y/N)                 "
1305 GOSUB 5500:IF Z$<>"Y" AND Z$<>"N" THEN 1305
1310 IF Z$="N" THEN RETURN


1315 REM ***************** ALTER DATA SUBROUTINE ******************
1320 GOSUB 5000:PRINT"LABEL NUMBER TO BE CHANGED":D2=0
1325 CL=32:LL=2:I=0:NO=0:GOSUB 6000:L=NX:IF L<J1 OR L>N THEN 1325
1330 RN=L-J1+4:CL=0:GOSUB 5000:PRINT SP$:GOSUB 5000:PRINT STR$(L);")"
1335 FOR K= 1 TO 3
1340 CL=8:I=0:LL=2:IF K>1 THEN I=1:LL=9:CL=K*12-8
1345 NO=C(L,K):GOSUB 6000:C(L,K)=NX:NEXT K:GOTO 1295
```

The following subroutines are required. (See Chapter 11)

4000 Clear screen routine.
4500 Home screen routine.
5000 Cursor position routine.
5500 Input character routine.
6000 Numeric input routine.

Important!
==========
Line 6035 to be changed as follows:

```
6035 IF Z$=RM$ AND D2=1 AND J>2 AND K=1 THEN 6100
```

9998 End program routine.

```
10000 DATA1,1000,0,2,1500,0,3,1500,0,4,2000,0
10005 DATA5,3000,0,6,3000,0,7,3000,0,8,11000,8000
10010 DATA9,4000,0,10,4000,0,11,5000,0,12,14000,52000
10015 DATA13,37000,0,14,5000,0,15,6000,0,16,6000,0
10020 DATA17,67000,88000,18,5000,0,19,5000,0
10025 DATA20,4000,0,21,59000,70000,22,3000,0,23,3000,0
10030 DATA24,2000,0,25,29000,75000,26,25000,0
10035 DATA27,4000,0,28,4000,0,29,3000,0,30,41000,48000
10040 DATA34,33500,46500,64,5500,22500
```

Chapter Eight

Contractor's Estimating

Estimating is a task which calls for a large quantity of basic information such as the cost of materials, rates of wages and anticipated output for a large number of building operations. In this chapter we show how this task can be adapted to a computer. To illustrate the principles we have chosen the trade of the bricklayer.

The program was written using as reference Chapter 7 of 'Estimating for Builders and Surveyors' by H. Enterkin and G. Reynolds, published by Heinemann.

The constants for labour and materials vary quite considerably in their units - e.g. bricks are purchased by the thousand, cement, lime and sand by the tonne. Labour output constants are usually based on the number of bricks per hour which can be laid and this will vary with the type of work involved. The number of bricks required to build a square metre of wall will depend on the wall thickness and also on the size of the bricks and of the mortar joints which are to be used. The final rates are quoted as the cost per cubic metre for mortar and the cost per square metre for brickwork. Extra over items are quoted as the cost per square metre or the cost per linear metre as appropriate.

The calculations which are involved in arriving at these rates are not difficult but they call on quite a number of labour and material constants and require quite a lot of arithmetic. To illustrate the calculations which are needed we give below the method used to obtain the cost per square metre of a half brick wall using common bricks in 1:3 cement mortar mixed by machine with two men operating the mixer. This is the first of the test examples for brickwork given later in this chapter. The prices and rates used are given in Table 8.1.

Stage 1 Calculate cost of materials for the mortar

Portland cement (delivered) /tonne	6700p
Labour unloading 1.25 tonnes/hour at 370p/hour	296p
	6996p

1 cu.m cement=1.44 tonnes @ 6996p/tonne	10074p
3 cu.m sand =4.56 tonnes @ 650p/tonne	2964p
	13038p
Allow 20% shrinkage (add 25%)	3260p
4 cubic metres cost	16298p
1 cubic metre costs	4075p

Stage 2 Calculate labour costs for the mortar

0.14/0.10 cu.m mixer hire rate 1500p per week	1500p	
1 labourer in charge 370+18=388p 40 h @ 388p	15520p	
1 labourer in attendance 40 h @ 370p	14800p	
	31820p	
Allow three hours standing time per week Cost per hour = 31820/37	860p	
Cleaning and maint. half hour per day of 8 h day, 1/16 of 388	24p	
1.14 litres of fuel @ 40p/litre	46p	
Oil and grease 670p/week, 630/37	18p	
	948p	
Output 1.15 cu.m/hour		
Cost of mixer per cu.m = 948/1.15		824p
Cost of 1 cu.m without profit		4899p

Stage 3 Calculate material costs for 1000 bricks

65 mm bricks delivered per 1000	14000p
Allow for waste 5%	700p
	14700p

Stage 4 Calculate labour costs to lay 1000 bricks

Based on two tradesmen and one labourer

2 tradesmen 1 h @ 428p	856p
1 labourer 1 h @ 370p	370p
	1226p

Output per hour = 55/tradesman

To lay 1000 bricks costs
1226 x 1000/110 11145p

Stage 5 Calculate mortar cost to lay 1000 bricks

1 cu.m of mortar costs 4899p (from stage 2)
0.53 cu.m mortar required/1000 bricks
Cost of mortar is 4899 x 0.53 2596p
28441p

Stage 6 Calculate rate per square metre

Assuming brick size is 215 x 65 mm
and joint is 10 mm thick
1 brick + joint = 0.225 x 0.075 m
Number of bricks per square metre
= 1/(0.225 x 0.075) = 59

Cost of 1 square metre = 28441 x 59/1000	1678p
Profit and oncost 20%	336p
	2014p

Program 8

Two methods of data input have been adopted. Firstly, standard brick and joint size and the productivity tables given in the above book have been put into DATA lines at the end of the program (lines 10005-10195). When the program is RUN the brick and joint sizes and the tables appear on the screen and an opportunity is given to change the sizes or productivity rates before proceeding with the remainder of the program.

A better way to store this information would be in a data file on disk but we have avoided the use of such files wherever possible because of the differences in creating and reading them on the various machines. If you decide to create a file for your use, the option to store any revised data can be given each time the program is run, otherwise if permanent changes are to be made it will be necessary to amend the data lines in the program.

You will notice that the text for the headings of the tables is put into the DATA lines in two forms. The texts for the first productivity table (10010-10050) are short enough to be accommodated on the minimum screen width of forty characters, still leaving room for the reference number at the beginning and the numeric values at the end. Spaces are included at the end of each text string to obtain a uniform length so that the numeric value at the end is tabulated without using the TAB statement. Some of the text for the other two tables is too long for a small screen and a different method of display has been used. No spaces are included at the end of the strings in this case (lines 10055-10100). Before productivity tables 2 and 3 are printed the numeric variable L4 is set to the maximum text length which can be accommodated. If the text exceeds this preset value, the sub-routine at line 2140 comes into play. This subroutine checks through the text string from the maximum possible length (L4) towards the beginning, looking for a space at which to split the string. The program then prints the first section followed by the remainder on the next line.

The second of the two numeric values attached to each table heading is not printed on the screen in Tables 1 and 2. In Table 1 these values are factors for the number of bricks which depends on wall thickness. For the other two tables the values are for wastage where applicable.

Costs of material and labour are entered each time the program is run. This could be changed to a disk data file, so that entries need only be made when changes are required. The remaining data lines (10105-10195) contain the prompts which are used when entering these values which is done in two sections using FOR-NEXT loops. The first section contains the costs of materials and the labour rates to be used and the second, the relevent data for a mixer. If hand mixing is chosen for the mortar this section of the program is bypassed. After each calculation an opportunity is given to amend any of the tables.

Table 8.1 Prices and rates to be used for testing purposes

Portland cement	6700 p/tonne
Snowcrete	14100 p/tonne
Lime	10000 p/tonne
Sand	650 p/tonne
Silver sand	1260 p/tonne
Common bricks	£140/1000
Facing bricks	£150/1000
Engineering bricks	£280/1000
Wall ties	£5.90/100
Labourer rate	370 p/hour
Tradesman rate	430 p/hour
Mixer hire rate	£15.00/week
Number of men on machine	2/1
Extra for labourer in charge	18 p/hour
Mixer output	1.15/0.59 cu.m/hour
Hours per week	40
Standing time	3 h/week
Fuel cost	40 p/litre
Fuel consumption	1.14 litres/h
Cleaning and maintenance	0.5 h/day
Oil and grease	670 p/week

So that you can check whether the estimating is done to your requirements we have listed below the various assumptions which have been made. To check the actual operation of the program we have given an example to test each of the possible options. To obtain the answers given in these examples, the prices and rates given in Table

8.1 must be used.

Assumptions

(a) Labourer unloads cement and lime at 1.25 tonnes/hour
(b) 1 cu.m of cement = 1.44 tonnes
(c) 1 cu.m of lime = 0.60 tonnes
(d) 1 cu.m of sand = 1.52 tonnes
(e) Shrinkage of cement mortar is 20%
(f) Shrinkage of cement/lime mortar is 25%
(g) Labourer takes 5 hours to hand mix 1 cu.m of mortar
(h) Wastage on common bricks is 5%
(j) 0.53 cu.m of mortar required/1000 bricks in half brick walls
(k) 0.60 cu.m of mortar required/1000 bricks in other walls
(l) Bricklaying team is two tradesmen and one labourer
(m) Fair face to common brickwork requires 5 minutes of labourer time/sq.m
(n) Labourer takes 2 hours to unload and stack facing or engineering bricks
(p) Cavity walls have 4 ties/sq.m

Test examples for mortar

1. Hand mixed 1:3	5924p/cu.m
2. Machine mixed 1:3 (2 men on machine)	4899p/cu.m
3. Machine mixed 1:3 (1 man on machine)	5003p/cu.m
4. Hand mixed 1:1:6	5547p/cu.m
5. Machine mixed 1:1:6 (2 men on machine)	4521p/cu.m
6. Machine mixed 1:1:6 (1 man on machine)	4625p/cu.m

Test examples for brickwork

In the following examples the mortar used is machine mixed cement mortar 1:3 with two men operating the mixer unless otherwise stated.

1. Half brick wall	£20.14/sq.m
2. One brick wall	£40.76/sq.m
3. One and a half brick wall	£59.17/sq.m
4. Reduced brickwork	£38.33/sq.m

5. Brickwork in projections £42.34/sq.m
6. Brickwork in backing to masonry £40.76/sq.m
7. Facing brickwork
   (a) half brick wall £24.63/sq.m
   (b) E/O stretcher bond, key pointed £7.27/sq.m
   (c) E/O english bond, point at later date £11.55/sq.m
   (d) E/O flemish bond, excluding pointing £5.69/sq.m
   (e) E/O garden wall bond, point at later date £10.47/sq.m
8. Engineering brickwork in manholes £64.25/sq.m
9. Engineering brick as facing brick
   (a) half brick wall £34.30/sq.m
   (b) E/O english bond, key pointed £24.01/sq.m
   (c) E/O flemish bond, excluding pointing £18.63/sq.m

Test examples for labours (commons)

1. E/O for rough arch two rings high, one brick soffit width £4.56/lin.m
2. E/O for key pointing as work proceeds £2.08/sq.m
3. E/O for rake out and point at later date £2.68/sq.m
4. E/O for form cavity £0.71/sq.m
5. E/O for rough cutting £6.98/sq.m

Test examples for labours (facings)

1. E/O fair cutting at vertical abutments
   (a) facing bricks £2.39/lin.m
   (b) engineering bricks £2.96/lin.m
2. E/O fair cutting at soffits (facings) £2.96/lin.m
3. E/O fair raking or splay cutting (eng.) £3.29/lin.m
4. E/O key or flush pointing £3.39/sq.m
5. E/O rake out and point using special mortar (1:3 hand mixed using snowcrete and silver sand) £5.37/sq.m

<u>Description of Program 8</u>

Lines 1020-1025 Dimension the necessary arrays. T$(), T() & F() hold the data fo table 1. L$() and L(,) hold the data fo

	table 2.
	L1$() and L1(,) hold the data for table 3.
	P$() holds the prompts used when pricing materials and entering labour rates.
	P() holds the values typed in for the above.
	M$() holds the prompts used when entering mixer costs.
	M() holds the values typed in for the above.
Lines 1030-1035	Initialise the general variables to be used in the program (see Chapter 11).
Lines 1050-1075	Read the data into the arrays.
Lines 1080-1090	Clear the screen and print the title.
Lines 1095-1110	Print the standard brick sizes.
Lines 1115-1120	Give the option to change the sizes.
Lines 1125-1135	Permit alterations to be made using the subroutine at line 6000.
Lines 1140-1165	Print the joint size and permit alteration in the same way as above for brick sizes.
Lines 1175-1210	Print Table 1 on the screen and permit alterations if required.
Lines 1220-1275	Do the same for Table 2.
Lines 1285-1360	Do the same for Table 3.
Lines 1370-1415	Allow the choice of mortar.
Lines 1425-1465	Require the input of material and labour costs on the first run through the program. On subsequent passes, the same section is used to amend any of these values.
Lines 1475-1525	Calculate the cost of the chosen mortar.
Lines 1535-1570	Require the input of mixer costs the first time a machine mixed mortar is chosen. Also used later to amend any of these values.
Lines 1580-1590	Calculate the cost of using the

	mixer.
Lines 1595-1600	Find the total cost of mortar and round the value to two decimal places.
Lines 1610-1635	Print the cost of mortar and options to calculate brickwork, labours or a different mortar. Option one prints Table 1 and permits the choice of the nine wall types (lines 1645-1660). Option two prints Table 2 (commons) or Table 3 (facings) as requested and permits the choice of the five labours (lines 1790-1830). Option three gives the option to change any prices, etc., and then returns to line 1370 (lines 1765-1780).
Lines 1645-1660	See option one under lines 1610-1635.
Line 1665	Sets the quantity of mortar required depending on the wall type chosen in option one.
Lines 1670-1720	Calculate brickwork costs.
Lines 1725-1735	Set prompt (P$) and round the result to two decimal places.
Lines 1745-1780	Display the result and ask if more calculations are required. If the answer is 'Y' the change price routine is encountered as described in option three under lines 1610-1635.
Lines 1790-1830	See option two under lines 1610-1635.
Lines 1835-1845	Branch to the section of program required for the labour chosen.
Lines 1850-1905	Special routines for the calculation of labours on commons.
Lines 1910-2055	Special routines for the calculation of labours on facings.

Subroutines

Lines 2065-2075	Check Y/N answers.
Lines 2085-2130	Permit the change of values in the various tables.
Lines 2140-2150	Divide text so that it will be printed on the screen without breaking a word at the end of the line.

The remaining subroutines are described in Chapter 11.

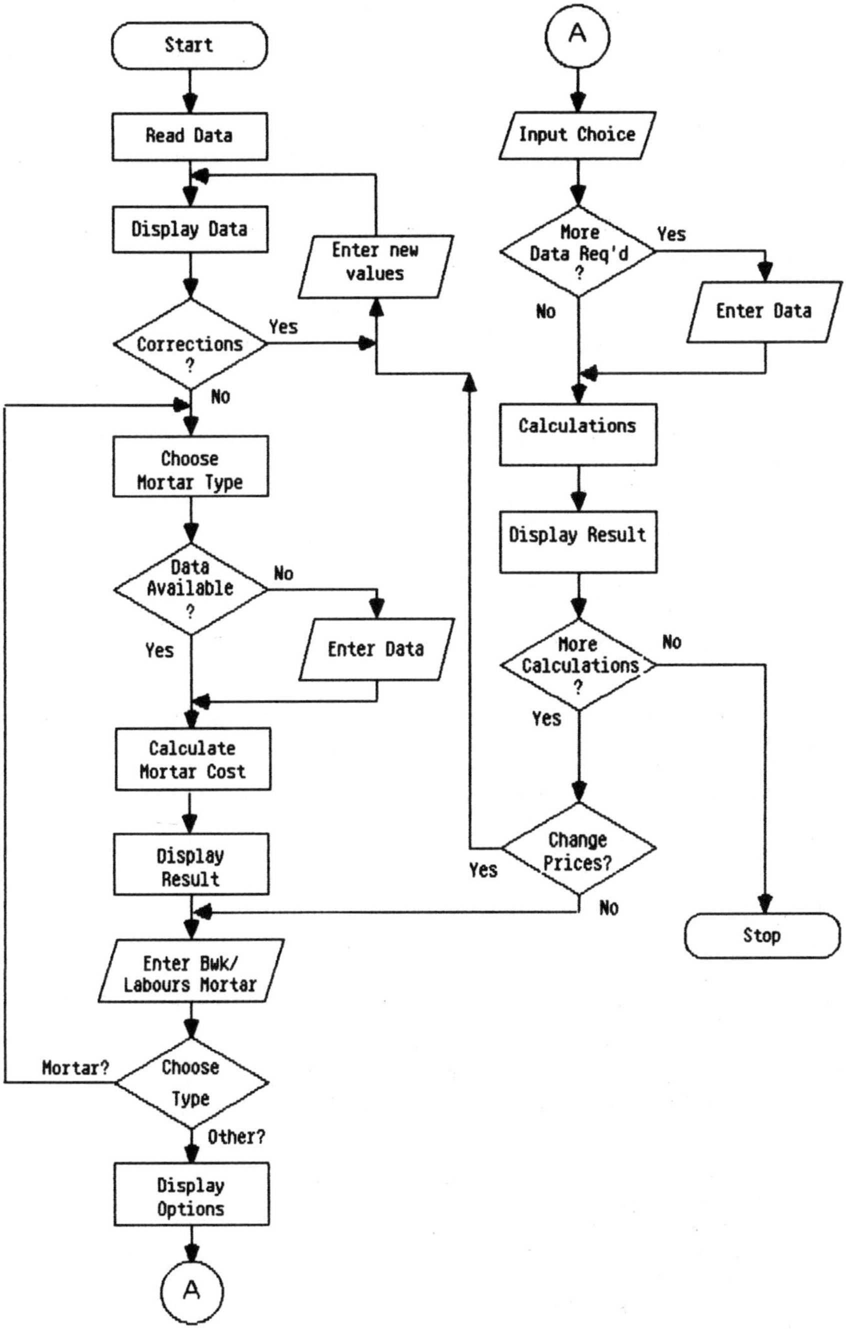

Figure 8.1 Flowchart for Program 8: Estimating

```
1000 REM **************** ESTIMATING. PROGRAM 8. *****************

1005 REM ************ COPYRIGHT BRANDON, MOORE & MAIN ************
1010 REM *COLLINS PROFESSIONAL & TECHNICAL BOOKS (C)84 REV8/12/84*

1015 REM ************ DIMENSION/INITIALISE VARIABLES *************
1020 DIM L(5,2),L$(5),L1(5,2),L1$(5),M$(10),M(10)
1025 DIM T$(9),T(9),P$(9),P(9),F(9)
1030 SP$="":FOR X= 1 TO 39:SP$=SP$+" ":NEXT X
1035 CS$=CHR$(147):HS$=CHR$(19):DEL$=CHR$(20):CE$="@":RM$="↑"
1040 D=0:D1=0:B$=SP$

1045 REM ***************** READ DATA INTO ARRAYS *****************
1050 READ L,W,H,J:REM                                 <READ DATA
1055 FOR X= 1 TO 9:READ T$(X),T(X),F(X):NEXT X
1060 FOR X= 1 TO 5:READ L$(X),L(X,1),L(X,2):NEXT X
1065 FOR X= 1 TO 5:READ L1$(X),L1(X,1),L1(X,2):NEXT X
1070 FOR X= 1 TO 9:READ P$(X):NEXT X
1075 FOR X= 1 TO 10:READ M$(X):NEXT X
1080 GOSUB 4000:REM                                CLEARS SCREEN
1085 PRINT LEFT$(SP$,9);"BRICKWORK ESTIMATING":PRINT

1090 REM ****************** BRICK & JOINT SIZE ******************
1095 PRINT"STANDARD BRICK SIZE IS-":PRINT
1100 PRINT"LENGTH";RIGHT$(SP$+STR$(L),8)"  MM":PRINT
1105 PRINT"WIDTH ";RIGHT$(SP$+STR$(W),8)"  MM":PRINT
1110 PRINT"HEIGHT";RIGHT$(SP$+STR$(H),8)"  MM":PRINT
1115 P$="DO YOU WISH TO CHANGE THESE?(Y/N)"
1120 GOSUB 2065:IF Z$="N" THEN 1140
1125 RN=5:CL=8:LL=6:I=1:NO=L:GOSUB 6000:L=NX
1130 RN=7:CL=8:LL=6:I=1:NO=W:GOSUB 6000:W=NX
1135 RN=9:CL=8:LL=6:I=1:NO=H:GOSUB 6000:H=NX
1140 RN=11:CL=0:GOSUB 5000
1145 PRINT"STANDARD JOINT SIZE IS-";RIGHT$(SP$+STR$(J),8);"  MM"
1150 P$="DO YOU WISH TO CHANGE IT?(Y/N)"
1155 GOSUB 2065:IF Z$="N" THEN 1165
1160 RN=11:CL=25:LL=6:I=1:NO=J:GOSUB 6000:J=NX
1165 N=1E6/((L+J)*(H+J)):N=INT(.5+N):P$="ANY CHANGES? (Y/N)"

1170 REM ************ AMEND PRODUCTIVITY DATA SCREEN 1 ***********
1175 GOSUB 4000:REM                                CLEARS SCREEN
1180 PRINT"PRODUCTIVITY (1)"
1185 PRINT"          TYPE OF WALL        OUTPUT/MAN HR":PRINT
1190 FOR X= 1 TO 9
1195 PRINT RIGHT$(SP$+STR$(X),2);")";T$(X);RIGHT$(SP$+STR$(T(X)),4)
1200 NEXT X
1205 GOSUB 2065:IF Z$="N" THEN 1220
1210 NN=9:D=0:D1=0:A1=1:GOSUB 2085:T(E)=NX:GOSUB 4500:GOTO 1180
```

```
1215 REM ************ AMEND PRODUCTIVITY DATA SCREEN 2 ***********
1220 GOSUB 4000:REM                                   CLEARS SCREEN
1225 PRINT"PRODUCTIVITY (2)":L4=25:REM              SET LINE LENGTH
1230 PRINT"          LABOURS             OUTPUT/MAN HR":PRINT
1235 FOR X= 1 TO 5
1240 PRINT RIGHT$(SP$+STR$(X),2);")";:L$=L$(X):PO=0:PO$=""
1245 IF LEN(L$)>L4-2 THEN GOSUB 2140:PRINT LEFT$(L$,PO):PO$="    "
1250 P=LEN(L$)-PO
1255 PRINT PO$;RIGHT$(L$,P);RIGHT$(SP$+STR$(L(X,1)),35-P-LEN(PO$));
1260 ME$="":IF X<>1 THEN ME$=" SQM"
1265 PRINT ME$:PRINT:NEXT X
1270 GOSUB 2065:IF Z$="N" THEN 1285
1275 NN=5:D=0:D1=0:A1=0:GOSUB 2085:L(E,1)=NX:GOSUB 4500:GOTO 1225


1280 REM ************ AMEND PRODUCTIVITY DATA SCREEN 3 ***********
1285 GOSUB 4000:REM                                   CLEARS SCREEN
1290 PRINT"PRODUCTIVITY (3)"
1295 PRINT"LABOURS ON FACING BRICKS   T/MAN BRICK"
1300 PRINT LEFT$(SP$,27);"HRS/  WASTE"
1305 PRINT LEFT$(SP$,27);"LIN.M /LIN.M"
1310 FOR X= 1 TO 5
1315 PRINT RIGHT$(SP$+STR$(X),2);")";:L$=L1$(X):PO=0:PO$=""
1320 IF LEN(L$)>L4-2 THEN GOSUB 2140:PRINT LEFT$(L$,PO):PO$="    "
1325 P=LEN(L$)-PO
1330 PRINT PO$;RIGHT$(L$,P);RIGHT$(SP$+STR$(L1(X,1)),32-P-LEN(PO$));
1335 IF X<4 THEN PRINT RIGHT$(SP$+STR$(L1(X,2)),7)
1340 IF X>3 THEN PRINT "   SQM."
1345 PRINT:NEXT X
1350 GOSUB 2065:IF Z$="N" THEN 1370
1355 NN=5:D=1:D1=0:A1=0:GOSUB 2085:L1(E,F)=NX:GOSUB 4500:GOTO 1290
1360 D4=0


1365 REM **********************************************************
1370 GOSUB 4000:REM                                   CLEARS SCREEN
1375 PRINT"MORTAR":PRINT:PRINT"IS THE MORTAR TO BE"
1380 PRINT:PRINT"1) HAND MIXED":PRINT"2) MACHINE MIXED"
1385 PRINT:PRINT"(1 OR 2)?"
1390 NN=2:D1=1:A1=1:GOSUB 2085:M=E
1395 RN=5:CL=0:GOSUB 5000
1400 PRINT LEFT$("1) 1:3"+SP$,20):PRINT LEFT$("2) 1:1:6"+SP$,20)
1405 NN=2:D1=1:A1=1:GOSUB 2085:G=E
1410 S=.25:IF G=2 THEN S=1/3
1415 IF P(1)<>0 AND D4=0 THEN 1505


1420 REM *********** INPUT/AMEND COSTS MATERIALS/LABOUR **********
1425 GOSUB 4000:REM                                   CLEARS SCREEN
1430 PRINT"COSTS OF MATERIALS D/D SITE & LABOUR":PRINT
1435 FOR X= 1 TO 9:A1=1:IF X>7 OR X<4 THEN A1=0
1440 PRINT RIGHT$(SP$+STR$(X),2);") ";LEFT$(P$(X)+SP$,28);
1445 IF D4=0 THEN 1465
```

```
1450 PRINT RIGHT$(SP$+STR$(P(X)),6):PRINT:NEXT X
1455 GOSUB 2065:IF Z$="N" THEN 1475
1460 NN=9:D=0:D1=0:A1=2:GOSUB 2085:P(E)=NX:GOSUB 4500:GOTO 1430
1465 RN=2*X+1:CL=32:LL=6:I=A1:NO=0:GOSUB 6000:P(X)=NX:PRINT:NEXT X


1470 REM ********************* CALCULATIONS *********************
1475 GOSUB 4000:REM                                    CLEARS SCREEN
1480 C1=P(1)+P(8)/1.25:REM                COST CEMENT/TONNE UNLOADED
1485 C1=C1*1.44:REM                                 COST CEMENT/CU.M.
1490 C3=P(3)*1.52:REM                                 COST SAND/CU.M.
1495 C2=P(2)+P(8)/1.25:REM                  COST LIME/TONNE UNLOADED
1500 C2=C2*.6:REM                                     COST LIME/CU.M.
1505 LO=0:IF G=2 THEN LO=C2
1510 C=C1+LO+C3*3*G:C=(C+S*C)/(4*G):REM          COST MATERIAL/CU.M.
1515 IF M=2 THEN 1525
1520 MO=C+5*P(8):IF M=1 THEN 1600
1525 IF M(1)<>0 AND D4=0 THEN 1580


1530 REM *********** INPUT/AMEND COSTS MATERIALS/LABOUR **********
1535 GOSUB 4000:REM                                    CLEARS SCREEN
1540 FOR X= 1 TO 10:A1=1:IF X=2 OR X=3 OR X=10 THEN A1=0
1545 PRINT RIGHT$(SP$+STR$(X),2);") ";LEFT$(M$(X)+SP$,28);
1550 IF D4=0 THEN 1570
1555 PRINT RIGHT$(SP$+STR$(M(X)),6):PRINT:NEXT X
1560 GOSUB 2065:IF Z$="N" THEN 1360
1565 NN=10:D=0:D1=0:A1=3:GOSUB 2085:M(E)=NX:GOSUB 4500:GOTO 1540
1570 RN=2*X-1:CL=32:I=A1:LL=6:NO=0:GOSUB 6000:M(X)=NX:PRINT:NEXT X


1575 REM ********************* CALCULATIONS *********************
1580 L3=P(8):IF M(2)=1 THEN L3=0
1585 MO=(M(1)*100+M(5)*(P(8)+M(3))+M(5)*L3)/(M(5)-M(6))
1590 MO=MO+(P(8)+M(3))/8*M(9)+M(8)*M(7)+M(10)/(M(5)-M(6))
1595 MO=MO/M(4)+C
1600 MO=INT(.5+MO)


1605 REM ******************* MENU OF OPTIONS ********************
1610 GOSUB 4000:REM                                    CLEARS SCREEN
1615 PRINT"COST OF MORTAR (WITHOUT PROFIT)=":PRINT MO;"PENCE/CU.M"
1620 PRINT:PRINT"DO YOU WISH TO ESTIMATE:-"
1625 PRINT:PRINT"1) BRICKWORK":PRINT:PRINT"2) LABOUR"
1630 PRINT:PRINT"3) MORTAR":PRINT:PRINT"(1,2 OR 3)?"
1635 NN=3:D1=1:A1=1:GOSUB 2085:ON E GOTO 1645,1790,1765


1640 REM ****************** ESTIMATE BRICKWORK ******************
1645 GOSUB 4000:REM                                    CLEARS SCREEN
1650 PRINT"CHOOSE WALL TYPE":PRINT
1655 FOR X= 1 TO 9:PRINT RIGHT$(SP$+STR$(X),2);")";T$(X):PRINT:NEXT X
1660 NN=9:D1=1:A1=1:GOSUB 2085:Q=E
1665 M2=.53:IF Q>1 THEN M2=.6:REM        SET QTY MORTAR/1000 BRICKS
1670 Q1=4:IF Q=7 THEN Q1=5
```

```
1675 IF Q>7 THEN Q1=6
1680 R1=(P(4)+.05*P(4))*100
1685 Z=P(Q1)*100+2*P(8):R2=Z+.05*Z:IF D5=1 THEN 1965
1690 IF Q<>7 AND Q<>9 THEN 1705
1695 RN=23:CL=0:GOSUB 5000:PRINT"HALF-BRICK WALL OR E/O (1 OR 2)?"
1700 NN=2:D=1:A1=1:GOSUB 2085:IF E=2 THEN 1935
1705 R0=R1:IF Q>6 THEN R0=R2
1710 S0=R0+1000/(2*T(Q))*(2*P(9)+P(8))
1715 S0=S0+M0*M2
1720 S0=S0/1000*N*F(Q)
1725 P$="RATE/SQ.M.  = "
1730 GOSUB 4000:REM                                   CLEARS SCREEN
1735 S0=S0+.2*S0:S0=INT(.5+S0)


1740 REM ***** DISPLAY RESULT/ ASK FOR MORE CALCS TO BE DONE *****
1745 GOSUB 4000:RN=4:CL=0:GOSUB 5000:PRINT P$;S0/100:PRINT
1750 PRINT"(INCLUDING 20% PROFIT & ON COST)":PRINT
1755 P$="ANY MORE CALCULATIONS (Y/N)?":GOSUB 2065
1760 IF Z$="N" THEN 9998
1765 P$="DO YOU WISH TO CHANGE ANY PRICES?(Y/N)":GOSUB 2065
1770 D3=0:D4=0:IF Z$="Y" THEN D4=1
1775 IF D4=0 THEN 1370
1780 M=0:GOTO 1425


1785 REM ******************** ESTIMATE LABOUR ********************
1790 RN=14:CL=0:GOSUB 5000:PRINT"COMMONS OR FACINGS (1 OR 2)?"
1795 NN=2:D1=1:A1=1:GOSUB 2085:ZZ=E
1800 GOSUB 4000:REM                                   CLEARS SCREEN
1805 PRINT"CHOOSE LABOUR REQUIRED":D3=1:L4=34:REM  SET LINE LENGTH
1810 PRINT:FOR X= 1 TO 5:PRINT RIGHT$(SP$+STR$(X),2);")";
1815 P0=0:P0$="":L$=L$(X):IF Z$="2" THEN L$=L1$(X)
1820 IF LEN(L$)>L4-2 THEN GOSUB 2140:PRINT LEFT$(L$,P0):P0$="    "
1825 PRINT P0$;RIGHT$(L$,LEN(L$)-P0):PRINT:NEXT X
1830 NN=5:D1=1:A1=1:GOSUB 2085:Q=E
1835 GOSUB 4000:REM                                   CLEARS SCREEN
1840 IF ZZ=2 THEN 1910
1845 ON Q GOTO 1850,1890,1895,1900,1905
1850 RN=2:CL=0:GOSUB 5000:PRINT"NUMBER OF RINGS"
1855 RN=2:CL=24:LL=3:I=0:N0=0:GOSUB 6000:R=NX
1860 RN=4:CL=0:GOSUB 5000:PRINT"SOFFIT WIDTH (NO. BKS)"
1865 RN=4:CL=24:LL=3:I=0:N0=0:GOSUB 6000:W0=NX
1870 Z=1000/(H+J)*R:Z=INT(.9+Z)
1875 S0=(Z/L(Q,1)-Z/T(2))*(P(9)+P(8)/2)
1880 S0=(S0+14/60*P(9))+2*(P(4)+.05*P(4))/10
1885 P$="EXTRA OVER/LIN. M. = ":GOTO 1735
1890 S0=P(9)/L(Q,1)+P(8)/12:GOTO 2055
1895 S0=P(9)/L(Q,1)+M0*.033*N/1000:GOTO 2055
1900 S0=P(7)/25*100+P(9)/L(Q,1):GOTO 2055
1905 S0=P(9)/L(Q,1)+L(Q,2)*P(4)/10:GOTO 2055
1910 IF Q>3 THEN 1970
1915 GOSUB 4000:REM                                   CLEARS SCREEN
1920 PRINT"TYPE OF BRICK:":PRINT:PRINT" 1) FACING."
1925 PRINT" 2) ENGINEERING.":PRINT:PRINT"(1 OR 2)?"
```

```
1930 NN=2:D1=1:A1=1:GOSUB 2085:Q1=E+4:D5=1:GOTO 1680
1935 GOSUB 4000:REM                                    CLEARS SCREEN
1940 PRINT"BOND:":PRINT:PRINT" 1) STRETCHER.":PRINT
1945 PRINT" 2) ENGLISH.":PRINT:PRINT" 3) FLEMISH.":PRINT
1950 PRINT" 4) GARDEN WALL":PRINT:PRINT"(1,2,3 OR 4)?"
1955 NN=4:D1=1:A1=1:GOSUB 2085:Q2=E:F=N:IF Q2=1 THEN 1970
1960 F=N+N/Q2:F=INT(.5+F)
1965 D5=0
1970 ON Q GOTO 1975,1975,1975,1980,1985,1985,1990,1990,1990
1975 S0=P(9)*L1(Q,1)+L1(Q,2)*R2/1000:GOTO 1885
1980 S0=P(9)/L1(Q,1)+P(8)/12:GOTO 2055
1985 S0=P(9)/L1(Q,1)+M0*.033*N/1000:GOTO 2055
1990 R0=R2-R1
1995 S0=1000/(2*T(Q))*(2*P(9)+P(8))-1000/(2*T(1))*(2*P(9)+P(8))
2000 S0=(S0+R0)*F/1000
2005 L4=38:RN=14:CL=0:GOSUB 5000
2010 FOR X= 4 TO 5:PRINT RIGHT$(SP$+STR$(X-3),2);")";
2015 P0=0:P0$="":L$=L1$(X)
2020 IF LEN(L$)>L4-2 THEN GOSUB 2140:PRINT LEFT$(L$,P0):P0$="    "
2025 PRINT P0$;RIGHT$(L$,LEN(L$)-P0):PRINT:NEXT X
2030 PRINT" 3) EXCLUDING POINTING":PRINT:PRINT"1,2 OR 3?"
2035 NN=3:D1=1:A1=1:GOSUB 2085:Z1=E+3
2040 IF Z1=6 THEN P$="EXCLUDING POINTING":GOTO 2050
2045 S0=S0+P(9)/L1(Z1,1):P$="INCLUDING"+L1$(Z1)
2050 GOSUB 4000:PRINT:PRINT P$:REM                     CLEARS SCREEN
2055 P$="EXTRA OVER/SQ.M. = ":GOTO 1735


2060 REM ******************** CHECK SUBROUTINE ******************
2065 RN=24:CL=0:GOSUB 5000:PRINT P$
2070 GOSUB 5500:IF Z$<>"Y" AND Z$<>"N" THEN 2070
2075 GOSUB 5000:PRINT SP$:RETURN


2080 REM **************** CHANGE VALUES SUBROUTINE ***************
2085 RN=24:CL=0:GOSUB 5000:PRINT"REF?"
2090 CL=5:LL=2:I=0:N0=0:GOSUB 6000:E=NX:IF E<1 OR E>NN THEN 2090
2095 IF D1=1 THEN CL=0:GOSUB 5000:PRINT SP$:RETURN
2100 F=1:IF D=0 OR E>3 THEN 2120
2105 RN=24:CL=0:GOSUB 5000:PRINT"ELEMENT?(1/2)"
2110 CL=14:LL=1:I=0:N0=0:GOSUB 6000:F=NX:IF F<1 OR F>2 THEN 2110
2115 F=NX
2120 RN=24:CL=19:GOSUB 5000:PRINT"NEW VALUE?"
2122 I=1:IF A1=2 AND (E>7 OR E<4) THEN I=0
2123 IF A1=3 AND (E=2 OR E=3 OR E=10) THEN I=0
2125 CL=30:LL=6:N0=0:IF A1=1 THEN I=0
2130 GOSUB 6000:RN=24:CL=0:GOSUB 5000:PRINT SP$:RETURN


2135 REM ******************* PRINT SUBROUTINE *******************
2140 FOR I1= L4 TO 1 STEP(-1)
2145 IF RIGHT$(LEFT$(L$,I1),1)=" " THEN P0=I1-1:I1=1
2150 NEXT I1:RETURN
```

The following subroutines are required. (See Chapter 11)

```
4000 Clear screen routine.
4500 Home  screen routine.
5000 Cursor position routine.
5500 Input character routine.
6000 Numeric input routine.
9998 End program routine.
```

```
10000 REM *********************** DATA ****************************
10005 DATA 215,102.5,65,10
10010 DATA" HALF-BRICK WALL                    ",55,1
10015 DATA" ONE-BRICK WALL                     ",55,2
10020 DATA" ONE AND HALF BRICK WALL            ",60,3
10025 DATA" REDUCED BRICKWORK                  ",65,2
10030 DATA" BRICKWORK IN PROJECTIONS           ",50,2
10035 DATA" BRICKWORK IN BACKING TO MASONRY",55,2
10040 DATA" FACING BRICKWORK                   ",40,1
10045 DATA" ENG. BRICKWORK IN MANHOLES         ",50,2
10050 DATA" ENG. BRICK AS FACING BRICK         ",40,1
10055 DATA" ROUGH ARCH (WASTE 12/SQM)",30,12
10060 DATA" KEY OR FLUSH POINTING AS WORK PROCEEDS",3,0
10065 DATA" RAKE OUT & POINT AT A LATER DATE",2,0
10070 DATA" FORM CAVITY IN HOLLOW WALLS (INCLUDING TIES)",12,0
10075 DATA" ROUGH CUTTING (WASTE 11/SQM)",1,11
10080 DATA" FAIR CUTTING AT VERTICAL ABUTMENTS",.33,3.5
10085 DATA" FAIR CUTTING AT SOFFITS",.5,2
10090 DATA" FAIR RAKING OR SPLAY CUTTING",.5,2
10095 DATA" KEY OR FLUSH POINTING AS WORK PROCEEDS",1.7,0
10100 DATA" RAKE OUT AND POINT AT A LATER DATE",1,0
10105 DATA"CEMENT (P/TONNE)?"
10110 DATA"LIME (P/TONNE)?"
10115 DATA"SAND (P/TONNE)?"
10120 DATA"COMMON BRICKS(POUNDS/1000)?"
10125 DATA"FACING BRICKS(POUNDS/1000)?"
10130 DATA"ENGINEERING BKS(POUNDS/1000)?"
10135 DATA"WALL TIES(POUNDS/100)?"
10140 DATA"LABOURER RATE (P/HOUR)?"
10145 DATA"TRADESMAN RATE (P/HOUR)?"
10150 DATA"MIXER HIRE RATE(POUNDS/WK)?"
10155 DATA"NUMBER OF MEN/MACHINE (1/2)?"
10160 DATA"EXTRA FOR LAB. I/C (P/HOUR)?"
10165 DATA"MIXER OUTPUT (CU.M/HOUR)?"
10170 DATA"HOURS/WEEK"
10175 DATA"STANDING TIME (HOURS/WEEK)?"
10180 DATA"FUEL COST (PENCE/LITRE)?"
10185 DATA"FUEL/HOUR (LITRES)?"
10190 DATA"CLEANING & MAINT (HOURS/DAY)?"
10195 DATA"OIL & GREASE (PENCE/WEEK)?"
```

Chapter Nine

Critical Path

Critical path or 'project network analysis' is a commonly used method for the planning and control of projects by the use of a diagrammatic representation. The diagram or flow chart shows the various tasks which are required to complete the project and the way in which they are inter-related. By estimating the time required for each individual task it is possible to calculate how long it will take to complete the project. The flow chart also shows the 'critical' tasks which govern the project length - a delay in one of these will lengthen the time taken to complete the work. Non-critical tasks can be seen to have a 'float' or permissible delay which can occur without affecting the finishing time.

The method has many advantages as it clearly shows the relationships between the various tasks - how one activity depends on or affects the others. As each task is allotted a time to be taken for completion it is relatively easy to see whether any of the planning is unrealistic. If the completion date has to be advanced it is quite clear which task or tasks must be speeded up in order to achieve the required result. In addition it may well be feasible to reduce the planned labour and plant requirement for some of the non-critical tasks, so making use of some of the float time.

For further reference on this subject see 'Building Research Station Digest' number 53

Procedure

The method can be used to represent a complete building project or some portion of it. The first step is always to decide what activities are involved and their sequence. This requires a good knowledge of the project and needs care to ensure that all activities required for successful completion have been included.

As an example we shall take the simple case of fitting out a room as a kitchen. The activities which will be involved will be

1. Floor tiling
2. Electrical carcase
3. Plumbing carcase
4. Carpenter fixing units
5. Electrical connections to appliances
6. Plumbing connections to appliances
7. Wall tiling
8. Decorating
9. Complete

The sequence of events will be in the order listed above but activities 2 and 3 can be proceeding at the same time as can activities 5, 6 and 7. This can be illustrated in diagrammatic form in two ways. Figure 9.1 is an arrow diagram in which each activity is represented by an arrow. The duration of the various activities is not considered at this stage.

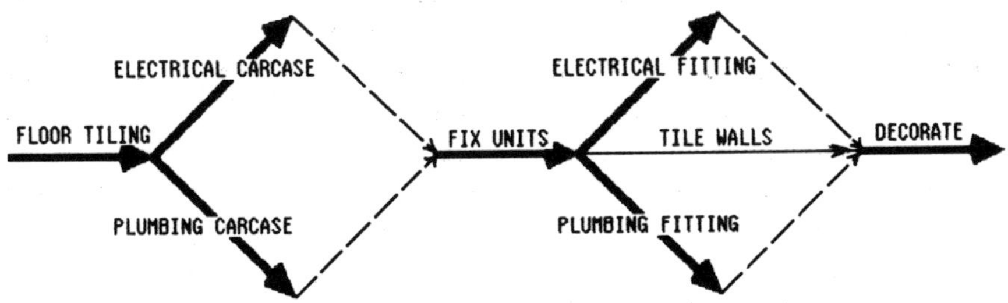

Figure 9.1 Arrow diagram

The broken arrows or 'dummy' arrows do not represent real activities but are put in to help to clarify the sequence of events. The second method avoids the use of dummy arrows, each activity being enclosed within a box as shown in Fig. 9.2.

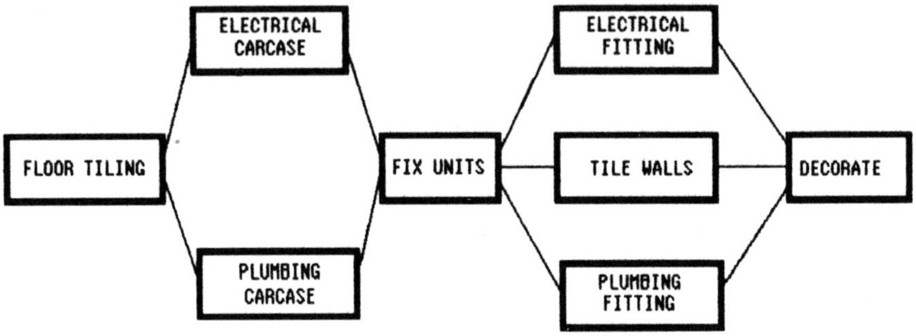

Figure 9.2 Box diagram

Now that the sequence of operations has been decided it is necessary to estimate the time which each operation will take. This can be in any convenient unit - in this case we shall use days.

Figure 9.3 repeats Fig. 9.2 but the estimated times for the various activities have been added. From this diagram we can see how soon an event can occur. No event can start until all those leading up to it have been completed, so that the earliest any event can occur is given by the longest alternative path to it. For example the fixing of the units (event number 4) cannot start until nine days after commencement of the job. Table 9.1 sets out the lengths of the alternative paths to each event, the longest of these appears in the third column of the table as the earliest time of the event in days from the commencement of the work.

Event number 9 is completion so that it can be seen from the table that this occurs sixteen days after the start and the path taken to arrive at this figure is through events 1, 3, 4, 7, 8, 9. This is called the 'Critical path'. This network is shown in Fig. 9.4.

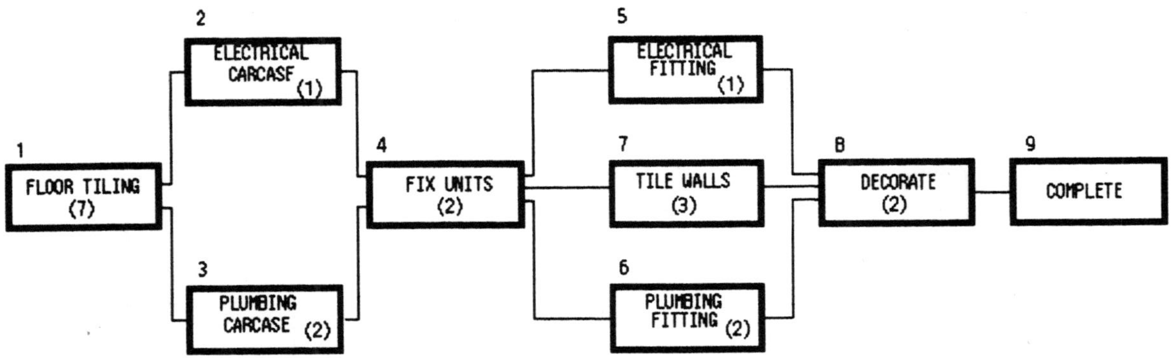

Figure 9.3 Box diagram showing estimated times in brackets

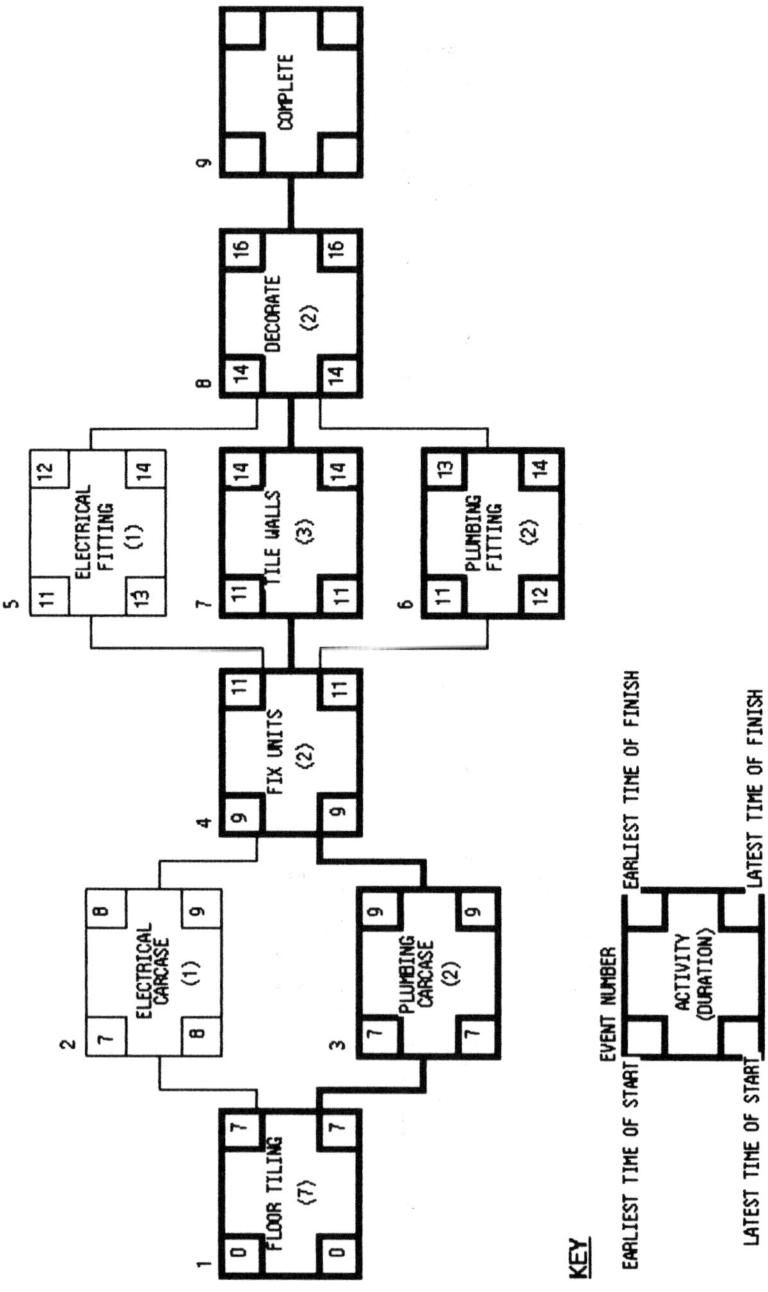

Figure 9.4 Completed network

Table 9.1

Event no.	Alternative paths	Earliest time of event (days after start)
1	-	0
2	7	7
3	7	7
4	7 + 1 or 7 + 2	9
5	9 + 2 + 2	11
6	9 + 2 + 2	11
7	9 + 2 + 2	11
8	11 + 1 or 11 + 3 or 11 + 2	14
9	14 + 2	16

In a simple example such as this it is easy to see the critical path by inspection but for a more complex scheme it is also necessary to calculate the latest time for each event - i.e. the maximum time that can be spent on a particular event without affecting the finishing time of the complete operation. The difference between the earliest and latest times an event can be completed is the float time and for critical tasks this will be zero. It follows that all tasks which are found to have the same earliest and latest times are critical. Therefore the critical path must pass through all these events.

Table 9.2

Event no.	Earliest time		Latest time
2	7	(9 - 1)	8
3	7	(9 - 2)	7
4	9	(11 - 2)	9
5	11	(14 - 1)	13
6	11	(14 - 2)	12
7	11	(14 - 3)	11
8	14	(16 - 2)	14
9	16		16

Table 9.2 shows how the latest times are calculated beginning at event number 9 and working backwards.

The critical path method is very useful in the planning stage of a job but it is also useful as a means of project control. The analysis can be undertaken at intervals during the progress of the work to determine the effect of what has actually happened and show what re-planning is required to the remaining activities in order to reduce the effect of any delays which may have occurred.

A full scale project may contain thousands of events but the method of analysis is the same. Although the arithmetic concerned is simple it becomes rather tedious especially if re-analysis is required at regular intervals and hence a computer program to do the arithmetic for you and indicate the critical path is a useful tool.

Description of Program 9

This program is designed to output the results to a printer so that it will be necessary to carry out a few modifications if you require the output on the screen. When using this program the option is available to enter a new project or to recall an old one which has been stored on disk. If a new project is entered, once the data have been checked, a file is created on the disk and the data are stored for future use. When an old project is recalled it is possible to add or delete tasks or to amend any of the existing ones. If this is done the new data are stored on the disk in the relevent file.

The entry of data routine requires the following information for each event:

1. Description of the event
2. I and J to indicate the sequence of events
3. Duration of the event

The I and J values can easily be obtained from the box diagram (Fig 9.2). We show a modified version of this in Fig 9.5 to which these values have been added.

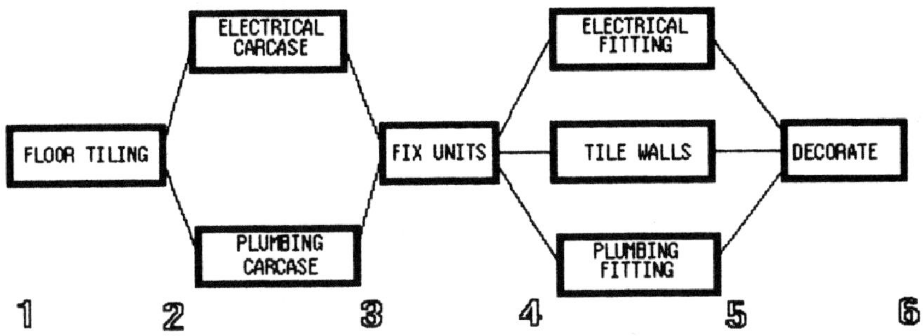

Figure 9.5 Box diagram showing I and J values to indicate sequence of events

For the first event (floor tiling) I=1 and J=2. For the second and third events which can be progressing simultaneously I=2 and J=3, etc.

The calculation section of the program (in lines 1330-1410) uses the Ford-Fulkerson algorithm to calculate the earliest and latest times that each event can occur. All events found to have the same earliest and latest completion times are critical as discussed earlier in this chapter.

Using the data for the example used in this chapter should produce the output shown in Table 9.3

Line 1020 Dimensions the array variables. The arrays (other than N$()) are dimensioned to 100 elements. A$() holds the description of the activities. I() and J() are the figures indicating the sequence of events. D() holds the duration of each event. I1() receives the earliest times of events for each different value in the array I() during the calculation process.

	I2() receives values which, when added to the overall time (I4), will give the latest finishing time for each event.
	N$() is used during the printing process.
Lines 1025-1035	Initialise the variables required in the program (see Chapter 11).
Lines 1045-1080	Print the headings and ask for the project number and also whether the project is new or old.
Lines 1090-1100	If an old project is to be recalled these lines read the information stored on the disk.
Lines 1105-1120	Ask for the project title and the evaluation date.
Lines 1125-1130	If the project is a new one these lines ask for the number of activities.
Lines 1140-1180	For an old project these lines display the first fifteen activities and for a new project allow the entry of the first fifteen activities.
Lines 1185-1195	Ask if any changes are to be made.
Lines 1200-1210	Print the editing menu (add, change or delete) and permit the choice of one of the three options.
Lines 1215-1255	These lines, together with lines 1280-1320 allow for editing to take place.
Lines 1260-1270	Increment the counter and check to see if more data are to be printed/input. If the check is positive, the screen is cleared of existing data and the program returns to line 1150.
Lines 1280-1295	Add activities (see above lines 1215-1255).
Lines 1300-1315	Delete activities (see above lines 1215-1255).
Lines 1325-1410	Clear the screen and display a mess-

	age to say that calculations are in progress. Calculation involves two nested loops - the outer loop R is set to a length of one more than the number of activities and the inner loop S is set to a length equal to the number of activities. During this process, the values are found for the arrays I1() and I2() and also the value of I4, which is the overall length of the project. If an error is detected during the calculations, the error flag (F) is set in line 1400. This causes the error message in line 1410 to be displayed.
Lines 1420-1595	Output the results to the printer and produce a table similar to that shown in table 9.3. Lines 1505 and 1510 put the I, J and duration values into I6, J1 and I7 and also calculate the required six values for the columns of the table (Y1 to Y6). Lines 1525 and 1530 transfer the above values into the array N() in the correct order for printing.
Lines 1605-1625	Store the data on a disk file.
Lines 1635-1665	Contain the subroutine which is called to input/amend an activity.

A description of the subroutines in the remainder of the program can be found in Chapter 11.

Table 9.3 Output from test example

Critical Path Program

Title: kitchen Project No: 1

Evaluation date: 21/1/85 No.of activies: 9

I	J	D	EST	LST	EFT	LFT	TOF	FRF	Activity	Type
1	2	7	0	0	7	7	0	0	floor tiles	Crit
2	3	1	7	8	8	9	1	1	electrical carcase	
2	3	2	7	7	9	9	0	0	plumbing carcase	Crit
3	4	2	9	9	11	11	0	0	fix units	Crit
4	5	1	11	13	12	14	2	2	electrical fitting	
4	5	2	11	12	13	14	1	1	plumbing fitting	
4	5	3	11	11	14	14	0	0	tile walls	Crit
5	6	2	14	14	16	16	0	0	decorate	Crit
6	7	0	16	16	16	16	0	0	complete	Crit

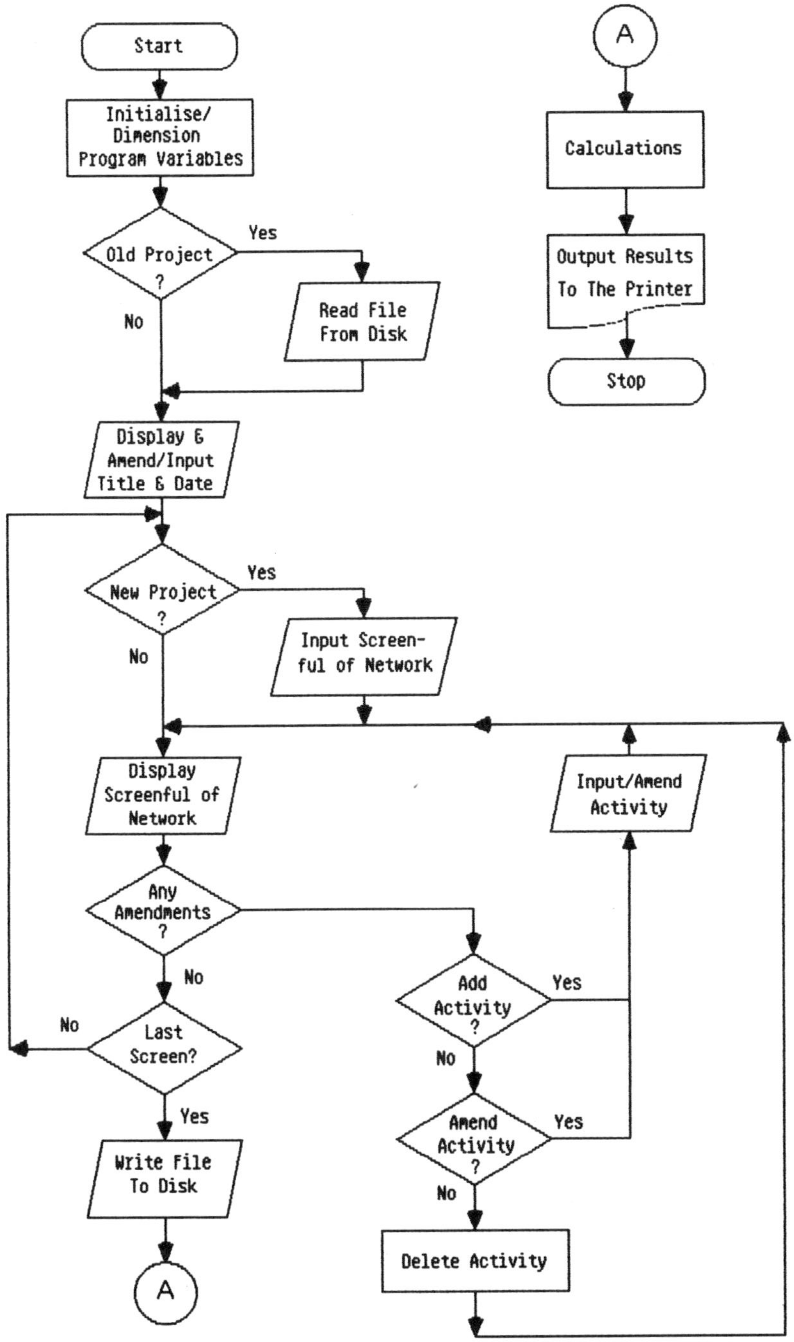

Figure 9.6 Flowchart for Program 9: Critical path

```
1000 REM***CRITICAL PATH PROGRAM 9. US'G FORD-FULKERSON ALGORITHM**

1005 REM************* COPYRIGHT BRANDON, MOORE & MAIN *************
1010 REM*COLLINS PROFESSIONAL & TECHNICAL BOOKS (C)84. REV21/2/85**

1015 REM*********** DIMENSION AND INITIALISE VARIABLES ************
1020 DIM A$(100),I(100),J(100),I1(100),I2(100),D(100),N(10),N$(11)
1025 CS$=CHR$(147):HS$=CHR$(19):DEL$=CHR$(20):RM$="↑":CE$="@"
1030 RV$="<":RO$=">":RT$=CHR$(13):FC$="!":LF$="":FL=0
1035 SP$="":LT$="":FOR X= 1 TO 39:SP$=SP$+" ":LT$=LT$+"-":NEXT

1040 REM**************** INPUT NEW/OLD PROJECT NUMBER **************
1045 GOSUB 4000:REM                                    CLEARS SCREEN
1050 PRINT LEFT$(SP$,10);RV$;"CRITICAL PATH PROGRAM";RO$:PRINT
1055 PRINT RV$;"O";RO$;"LD OR ";RV$;"N";RO$;"EW PROJECT"
1060 GOSUB 5500:P$=Z$:IF P$<>"O" AND P$<>"N" THEN 1060
1065 RN=6:CL=0:Q$="OLD JOB ":IF P$="N" THEN Q$="NEW JOB "
1070 QU$=Q$+"NUMBER":LL=3:C$="":GOSUB 9500:GOSUB 7000:PN$=C$
1075 DF$="CRIT "+RIGHT$(SP$+PN$,3)
1080 IF P$="N" THEN 1105

1085 REM************* INPUT OLD PROJECT DATA FROM DISK ************
1090 DOPEN#1,(DF$),D0
1095 INPUT#1,N,N$,ED$
1100 FOR X= 1 TO N:INPUT#1,A$(X),I(X),J(X),D(X):NEXT X:DCLOSE#1
1105 RN=9:CL=0:QU$=Q$+"TITLE ":LL=20:C$=N$:GOSUB 9500
1110 GOSUB 7000:N$=C$
1115 RN=12:CL=0:QU$="EVALUATION DATE   ":LL=8:C$=ED$:GOSUB 9500
1120 GOSUB 7000:ED$=C$:IF P$="O" THEN 1140
1125 RN=15:CL=0:QU$="NO. OF ACTIVITIES ":LL=2:NO=0:GOSUB 9500
1130 GOSUB 6000:N=NX:IF N>97 THEN 1125

1135 REM***************** DISPLAY/INPUT ACTIVITY *****************
1140 GOSUB 4000:REM                                    CLEARS SCREEN
1145 PRINT"NO       ACTIVITY            I    J   DURN":PRINT LT$:J1=1
1150 RN=2:FOR X= J1 TO J1+15:RN=RN+1:CL=0:GOSUB 5000
1155 PRINT RIGHT$(SP$+STR$(X),2);" ";LEFT$(A$(X)+SP$,20);
1160 PRINT RIGHT$(SP$+STR$(I(X)),4);" ";RIGHT$(SP$+STR$(J(X)),4);
1165 PRINT"  ";RIGHT$(SP$+STR$(D(X)),4)
1170 IF P$="N" AND FL=0 THEN L=X:GOSUB 1635
1175 IF X=N THEN PRINT SP$:GOTO 1185
1180 NEXT X:PRINT SP$
1185 RN=24:CL=0:GOSUB 5000:PRINT LEFT$("ANY CHANGES? (Y/N)"+SP$,39)
1190 GOSUB 5500:IF Z$<>"Y" AND Z$<>"N" THEN 1190
1195 IF Z$="N" THEN GOSUB 5000:PRINT SP$:GOTO 1260
1200 GOSUB 5000:PRINT RV$;"A";RO$;"DD,";RV$;"C";RO$;"HANGE OR ";
1205 PRINT RV$;"D";RO$;"ELETE?"
1210 GOSUB 5500:B$=Z$:IF B$<>"A" AND B$<>"C" AND B$<>"D" THEN 1210
```

```
1215 Q$="ADD AFTER":IF B$="D" THEN Q$="DELETE"
1220 IF B$="C" THEN Q$="CHANGE"
1225 RN=24:CL=0:GOSUB 5000:PRINT LEFT$(Q$+" NUMBER"+SP$,39)
1230 RN=24:CL=18:LL=2:I=0:NO=0:GOSUB 6000:L=NX
1235 IF L<J1 OR L>(J1+15) OR L>N THEN 1255
1240 IF B$="C" THEN RN=L-J1+3:GOSUB 1635:FL=1:GOTO 1150
1245 IF B$="A" THEN 1280
1250 IF B$="D" AND N>2 THEN 1300
1255 ER$="ILLEGAL!":QU$="":QU=18:GOSUB 7505:GOTO 1185
1260 J1=J1+16:IF J1=97 OR J1>N THEN GOSUB 1605:GOTO 1325
1265 RN=2:CL=0:GOSUB 5000:FOR Z= 1 TO 17:PRINT SP$:NEXT Z
1270 FL=0:GOTO 1150


1275 REM****************** ADD ACTIVITY OPTION ********************
1280 N=N+1:L=L+1:IF L=N THEN 1295
1285 FOR X= N TO L STEP(-1):U=X-1
1290 A$(X)=A$(U):I(X)=I(U):J(X)=J(U):D(X)=D(U):NEXT X
1295 A$(L)="":I(L)=0:J(L)=0:D(L)=0:RN=24:GOSUB 1635:FL=1:GOTO 1150
1300 IF L=N THEN 1315
1305 FOR X= L TO N:U=X+1
1310 A$(X)=A$(U):I(X)=I(U):J(X)=J(U):D(X)=D(U):NEXT X
1315 A$(N)="":I(N)=0:J(N)=0:D(N)=0:N=N-1:FL=1:GOTO 1150


1320 REM*********************** CALCULATIONS ***********************
1325 GOSUB 4000:REM                                     CLEARS SCREEN
1330 PRINT"PLEASE WAIT WHILE THE CALCULATIONS ARE  MADE..."
1335 F=0:M=N:I3=N+1:I4=0
1340 FOR R= 1 TO M:I1(R)=0:I2(R)=0:NEXT R
1345 FOR R= 1 TO I3:I5=0
1350 FOR S= 1 TO M:I6=I(S):J1=J(S):I7=D(S)
1355 ON (SGN(I1(J1)-I1(I6)-I7)+2) GOTO 1360,1380,1380
1360 I1(J1)=I1(I6)+I7
1365 ON (SGN(I1(J1)-I4)+2) GOTO 1375,1375,1370
1370 I4=I1(J1)
1375 I5=1
1380 ON (SGN(I2(I6)-I2(J1)+I7)+2) GOTO 1390,1390,1385
1385 I2(I6)=I2(J1)-I7:I5=1
1390 NEXT S
1395 ON (SGN(I5)+2) GOTO 1405,1400,1405
1400 R=I3:F=1
1405 NEXT R:IF F=1 THEN 1420
1410 PRINT:PRINT"ERROR IN DATA, TRY AGAIN":GOSUB 9000:GOTO 1140


1415 REM************** OUTPUT RESULTS TO THE PRINTER **************
1420 F=0:N1=1
1425 OPEN2,4:Z$=LF$:REM                         OPEN PRINTER CHANNEL!!!
1430 PRINT#2,LEFT$(SP$,30)"CRITICAL PATH PROGRAM";Z$
1435 PRINT#2,LEFT$(SP$,30);LEFT$(LT$,21);Z$
1440 PRINT#2,"TITLE: ";LEFT$(N$+SP$,25);"PROJECT NO: ";PN$;Z$
1445 PRINT#2,Z$:PRINT#2,"EVALUATION DATE: ";LEFT$(ED$+SP$,15);
1450 PRINT#2,"NO.OF ACTIVIES:";N;Z$:PRINT#2,Z$
1455 T$="!  I    J    D  ! EST   LST   EFT   LFT ! TOF   FRF !"
```

```
1460 T$=T$+"      ACTIVITY      !TYPE !"
1465 U$="!"+LEFT$(LT$,15)+"!"+LEFT$(LT$,23)+"!"+LEFT$(LT$,11)+"!"
1470 U$=U$+LEFT$(LT$,19)+"!"+LEFT$(LT$,5)+"!"
1475 S$="!"+LEFT$(SP$,15)+"!"+LEFT$(SP$,23)+"!"+LEFT$(SP$,11)+"!"
1480 S$=S$+LEFT$(SP$,19)+"!"+LEFT$(SP$,5)+"!"
1485 FO$=CHR$(36)+CHR$(36)+CHR$(69)+CHR$(37)+CHR$(37)+CHR$(37)
1490 FO$=FO$+CHR$(69)+CHR$(37)+CHR$(69)+CHR$(146)+CHR$(132)
1495 PRINT#2,U$;Z$:PRINT#2,T$;Z$:PRINT#2,U$;Z$
1500 FOR L= 1 TO 150
1505 I6=I(N1):J1=J(N1):I7=D(N1):Y1=I1(I6):Y2=Y1+I7:Y3=I2(J1)+I4
1510 Y4=Y3-I7:Y5=Y3-Y1-I7:Y6=I1(J1)-Y1-I7
1515 W$="    ":IF Y5=<0 THEN W$="CRIT"
1520 PRINT#2,S$;Z$
1525 N(1)=I6:N(2)=J1:N(3)=I7:N(4)=Y1:N(5)=Y4:N(6)=Y2:N(7)=Y3
1530 N(8)=Y5:N(9)=Y6:N$(10)=A$(N1):N$(11)=W$:GOSUB 8000:N1=N1+1
1535 ON (SGN(N1-M)+2) GOTO 1545,1545,1540
1540 L=150
1545 NEXT L
1550 PRINT#2,S$;Z$:PRINT#2,U$;Z$:PRINT#2,Z$
1555 PRINT#2,"KEY";Z$:PRINT#2,"---";Z$
1560 PRINT#2," D=DURATION BETWEEN EVENT 'I'AND EVENT 'J'";Z$
1565 PRINT#2,"EST=EARLIEST STARTING TIME";Z$
1570 PRINT#2,"LST=LATEST STARTING TIME";Z$
1575 PRINT#2,"EFT=EARLIEST FINISHING TIME";Z$
1580 PRINT#2,"LFT=LATEST FINISHING TIME";Z$
1585 PRINT#2,"TOF=TOTAL FLOAT";Z$
1590 PRINT#2,"FRF=FREE FLOAT";Z$
1595 CLOSE2:GOTO 9998:REM                    CLOSE PRINTER CHANNEL!!!


1600 REM************* WRITE DATA TO DISK FILE ROUTINE *************
1605 SCRATCH(DF$):DOPEN#1,(DF$),D0,W
1610 PRINT#1,N;RT$;N$;RT$;ED$;RT$;
1615 FOR X= 1 TO N
1620 PRINT#1,A$(X);RT$;I(X);RT$;J(X);RT$;D(X);RT$;
1625 NEXT X:DCLOSE#1:RETURN


1630 REM************* INPUT/AMEND ACTIVITY SUBROUTINE *************
1635 CL=0:GOSUB 5000:PRINT RIGHT$(SP$+STR$(L),2);" "
1640 CL=3:LL=18:C$=A$(L):GOSUB 7000:A$(L)=C$
1645 CL=25:LL=2:NO=I(L):GOSUB 6000:I(L)=NX
1650 CL=30:LL=2:NO=J(L):GOSUB 6000:J(L)=NX
1655 CL=35:LL=3:NO=D(L):GOSUB 6000:D(L)=NX
1660 IF I(L)>(N+1) OR J(L)>(N+1) THEN 1645
1665 RETURN
```

The following subroutines are required. (See Chapter 11)

4000 Clear screen rtn.
4500 Home screen rtn.
5000 Cursor position rtn.
5500 Input character rtn.
6000 Numeric input rtn.
7000 Alphanumeric input rtn.
7500 Report error rtn.
8000 Format/print line rtn.
9000 Pause rtn.
9500 Draw input box rtn.
9998 End program rtn.

Chapter Ten

Monte Carlo Simulation

Until now we have assumed in most of the techniques that the values we give to the variables in the formulae are certain. However in the real world we know that it is almost impossible to predict the life of a building, the cost of a component, the internal rate of return of a company or whatever. These values are subject to uncertainty, relying as they do on human behaviour. However it is often possible to identify a distribution for these variables from which the real value is likely to come. If we sample from this distribution to obtain the figures which we then place in our formula we can model the uncertainty that exists in reality. By repeating the calculations a number of times, again sampling from the distributions, a distribution of the final results can be obtained which reflects the probability of a certain cost occurring.

One method of sampling from the distribution is known as 'Monte Carlo simulation'. It is so called because it uses random numbers, not unlike a dice throw or roulette wheel, to select by chance a value from the distribution. The technique is very useful because it allows the simulation of activities over time, of a kind which cannot be represented by any of the theoretical models. Simulation in this context simply means the creating of a typical life history of the system and its activities (e.g. building process, cost-in-use, etc.) under given conditions, working out step-by-step what happens during each unit of life of the system.

Example

The example chosen for the program in this book concerns a building component which requires no routine maintenance or repair but which normally needs to be replaced several times during the building life. To avoid making the problem too complex for the educational purposes of this book, simplifying assumptions have been made and taxation, inflation, disruption costs and secondary maintenance have not been included. It would however be easy to add these should they be required. The variables chosen for this simple simulation are:

(i) Component life
(ii) Life of building
(iii) Interest rate

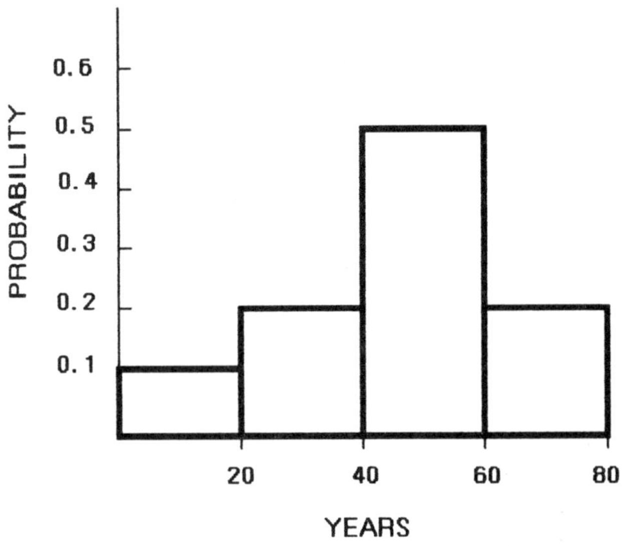

Figure 10.1 Histogram of building life

Hypothetical distributions for each of these have been prepared and the technique for entering these into the program will be illustrated with the variable 'building life'. Figure 10.1 shows a histogram for the probable life expectancy of the building. Thus there is a probability of 0.1 that the building life will be less than twenty years and a probability of 0.5 that it will be somewhere between forty and sixty years. In reality you may feel that there is no probability that the building will be pulled down within five years and therefore the distribution should commence at this point on the scale. The fact that the distribution is not symmetrical makes it more difficult to describe by one of the common distributions and it is therefore necessary to construct a cumulative or ogive for each variable. To translate the histogram into a cumulative distribution it is a simple matter of adding each probability to the previous one, in order, starting with the lowest class interval. This is shown in the following table.

Building life	Probability	Cumulative building life	Cumulative building life
0-20	0.1	0-20	0.1
20-40	0.2	0-40	0.3
40-60	0.5	0-60	0.8
60-80	0.2	0-80	1.0

These points can then be plotted on a graph as shown in Fig. 10.2 for all these distributions. To sample from the distribution it is now a simple matter of picking a random number (generated by the computer or taken from tables) and reading off the resultant life of the building. Notice that the steeper the slope of the ogive the wider the range of random numbers that will select this portion of the total curve and therefore the higher the probability of this class interval being selected. One word of caution should be expressed concerning the use of random numbers generated by a computer. In practice they are usually pseudo random numbers and if the operation of selection is performed too many times the same sequence of numbers will begin to appear again.

Having established the principles it is now merely a question of working out the order of sampling and calculation. For our simple component it is a matter of:

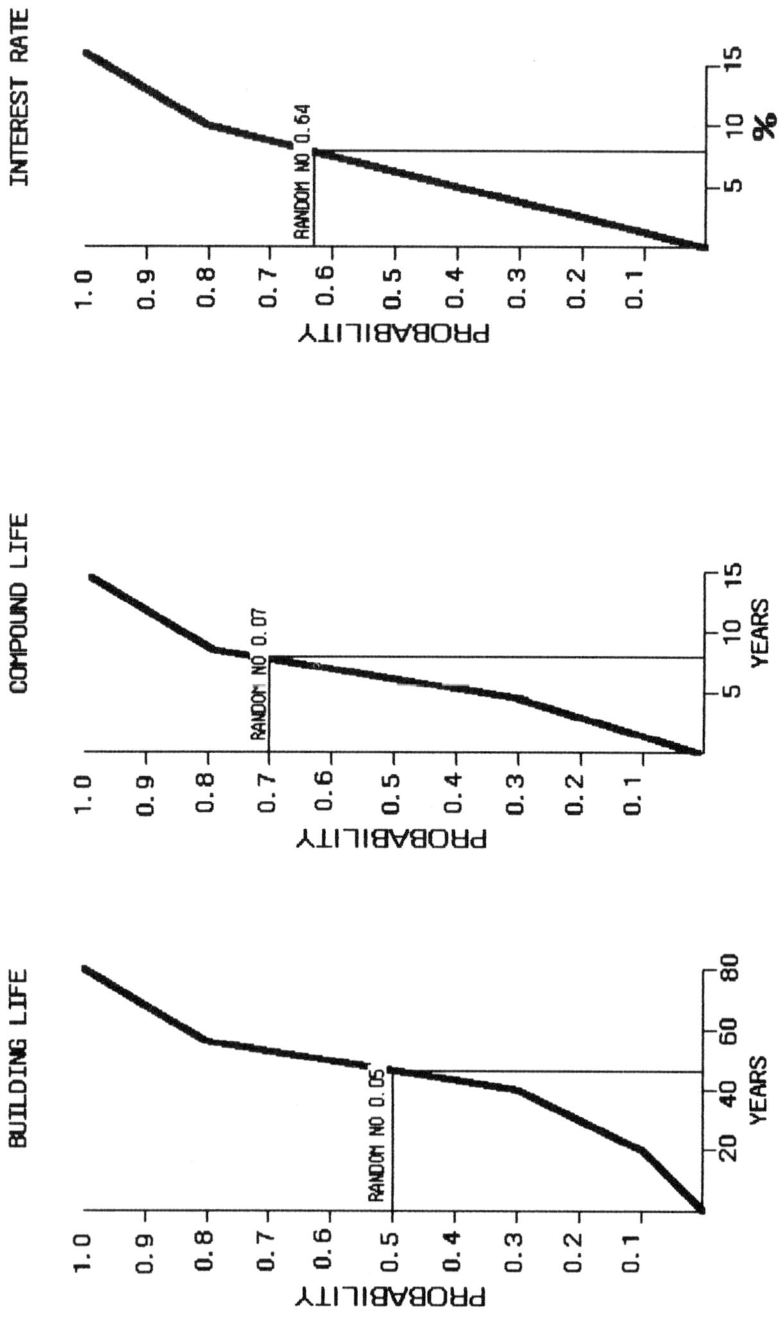

Figure 10.2 Cumulative distribution curves

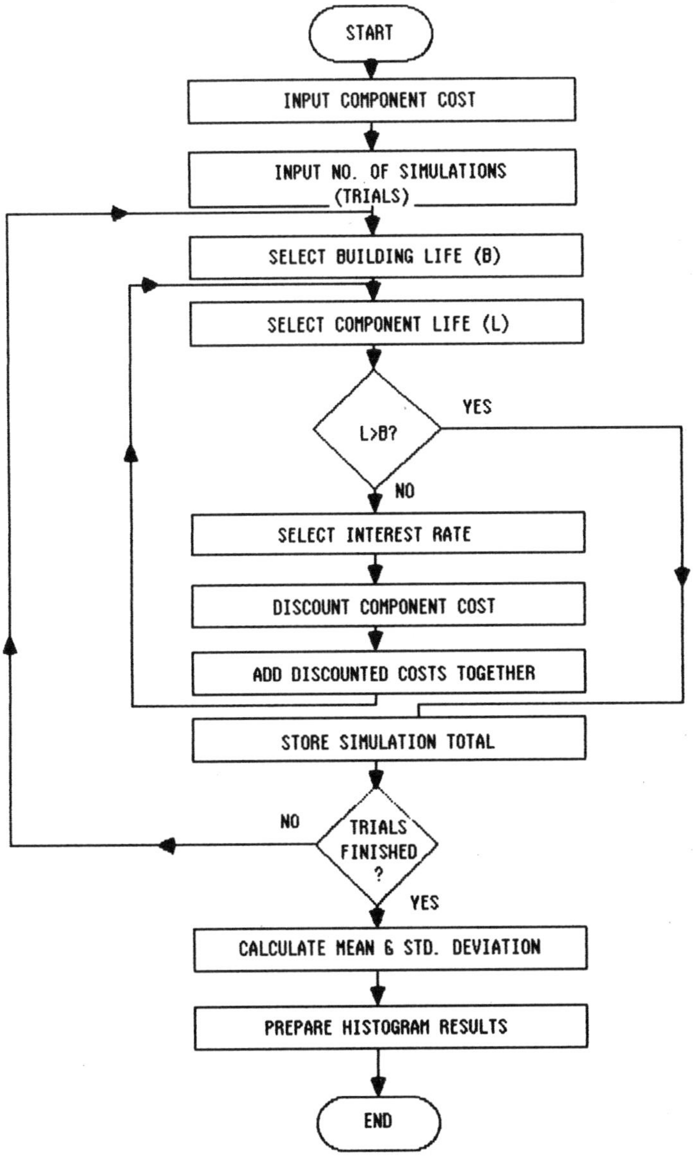

Figure 10.3 Flow chart Monte Carlo CIU model

1. Choosing the building life.
2. Selecting the sequence of component lives from the relevent distribution until their sum equals or exceeds the building life.
3. Discounting the cost of the replacement of the component using a discount rate selected from the interest rate distribution.
4. Adding together the initial cost and sum of the discounted value of the future costs and storing the result.
5. Forming a distribution from the collection of results obtained from the specified number of trials.

For a computer simulation it is more convenient to follow the procedure shown in the flow chart in Fig. 10.3. The following steps show this procedure in relation to the program.

Line 1015	Initialises two variables. The first is a number seed used to generate a different random number pattern on each call of the RND function. (See subroutine 1435.) The second, variable 'NL', is the number of columns that the user's computer is capable of displaying. This is used in the histogram routine to make the best possible use of the display.
Line 1020	Dimensions the array which holds the results of the trials and also the array which holds the frequencies.
Lines 1025-1080	Titles followed by selection of initial cost of component.
Lines 1085-1090	Select the number of trials (iterations) required. For machines with memories of less than 32k, it may be necessary to reduce the maximum number allowable.
Lines 1100-1110	Initialise the control loop and display the number of completed trials as the calculations progress.
Lines 1115-1140	Select building life (B). The mach-

	ine asks if the random number is greater than 8 (the random number has been multiplied by 10). If so it then calculates the building life from the formula given. If not it goes on and tries the other class intervals.
Lines 1145-1165	Sample from the distribution of component life in a similar way to above.
Lines 1170-1185	Sample from the distribution of interest rates as above.
Lines 1190-1210	Calculate discounted values of replacements for component life (note 10% has been added for extra cost for replacements over and above initial cost). This is an iterative process returning to selection of component life until the building life has been equalled or exceeded by the total of the component lives. A running total is kept for each trial and when the trial is complete the total is entered into store (T) for future analysis, i.e. arithmetic means and standard deviations.
Lines 1215-1245	Calculate arithmetic mean and standard deviation from results held in array T.
Lines 1250-1265	Calculate and display a histogram of the results.
Lines 1270-1285	Give the opportunity to run another simulation.
Lines 1290-1430	Show a summary of arithmetic mean and standard deviation.

Note that the selection of a random number is made the subject of a subroutine in line 1435 and is used each time a random number is required.

Providing the calculation of cost-in-use and the sampling routines are understood then the program should be reasonably easy to follow except perhaps for the simple histogram that is produced on the screen. This has been made more complex than would normally be required to cope with a variety of present value totals. The procedure is as follows:

The first step is to find the maxima and minima of the data to be displayed, then to scale the information so that it fits on the screen.

Lines 1295-1315	Find the maximum and minimum costs that have been generated by the computer. Initially the minimum 'start' value (variable MI) has to be set to a very large figure and the maximum 'start' value (variable MA) to zero. It then examines each element in the result array T() to see if it is greater than the maximum and/or less than the minimum. The value is stored as the new maximum/minimum if either test proves true.
Line 1320	The next step is to expand the range of the maximum/minimum values by increasing each by 5%. The interval is then found by finding the difference and dividing by 10. (Variable RA)
Lines 1325-1340	The result array can now be transposed into another array S() which finds how many results fall within each of the intervals.
Lines 1345-1355	The maximum number of results which fall within any one interval must now be found to find the horizontal scaling factor. The maximum is placed in variable HI.
Line 1360	This maximum must now be divided by the number of characters available for the 'bar' of the histogram which in the case of a forty column machine is 25. Thus the frequency

factor can be calculated and placed in FQ.

Lines 1370-1425 The display of the histogram is achieved by displaying the value of each interval (starting from the minimum and adding an interval at each successive step) followed by the number of characters which represent the frequency of that interval. This is found by dividing the interval frequency by the frequency factor FQ.

Results

It is not possible to accurately predict the results each machine will give as it depends on the random number sequence. However a CBM 8032 gave the output shown in Fig. 10.4 when an initial cost of one hundred pounds and two hundred trials were entered.

```
HISTOGRAM CIU SIMULATION(MONTE CARLO)

EACH * REPRESENTS A FREQUENCY OF 2

--------------------------------------
        95  ******
    143.66  **********
    192.32  *********************
    240.98  *********************************
    289.64  *****************
     338.3  *******
    386.96  **
    435.62  *
    484.28  *
    532.94
INITIAL COST      = 100 POUNDS
ARITHMETIC MEAN   = 261.18 POUNDS
STANDARD DEVIATION= 273.08 POUNDS
--------------------------------------
DO YOU WISH TO TRY AGAIN? (Y/N)
```

Figure 10.4 Output of results

Conclusion

This particular program is illustrative of the manner by which the Monte Carlo simulation technique can be computerised. For simplicity, subroutines have been kept to a minimum. For a practical program which can be used for a number of problems and where the data or distributions, etc., were fed in through the keyboard then subroutines for each operation would be the sensible way to handle the organisation of the program.

The program can be improved in any number of ways but if the purpose is merely to learn the operation of the technique then we suggest that the reader attempts to include the following.

1. Allowance for irregular maintenance costs by selecting from a further distribution.
2. Incorporation of inflation rate into the future cost calculation. This raises a number of problems, including the question of what the distribution for inflation should be and how the rate should be linked to the interest rate.

You may also like to try and solve the problem of how do you reflect a realistic change in interest rates over time as in the present program it is quite possible for the interest rate to fluctuate between the minimum and maximum in just one time period. One approach could be to generate the interest rate for every year of the building life and place in a separate array right at the beginning of each iteration - ready to be used when required. A constraint could be imposed to ensure that each subsequent year's interest rate as it is being generated does not exceed an increase or decrease of say 20% of the previous year's interest rate.

The purpose of this chapter has been to provide a launching pad for more ambitious programs that the reader may wish to develop.

```
1000 REM********** MONTE CARLO SIMULATION. PROGRAM 10. ***********

1005 REM******* COPYRIGHT BRANDON, MOORE AND MAIN (C) 1984 *******
1010 REM*COLLINS PROFESSIONAL & TECHNICAL BOOKS C(84). REVN6/3/85*
1015 Z9=.1234:NL=40:REM ***   Z9=SEED RANDOM NO. NL=NO. OF COLUMNS
1020 DIM T(600),S(20):REM ***                     DIMENSION ARRAYS
1025 PRINT CHR$(147);:REM                          CLEARS SCREEN!!!
1030 PRINT"    *** MONTE CARLO SIMULATION CIU ***":PRINT
1035 PRINT"THIS  PROGRAM  DEMONSTRATES  THE USE OF"
1040 PRINT"MONTE CARLO TECHNIQUE  FOR CIU  MODELS."
1045 PRINT"IT  SELECTS THE LIFE OF  THE  BUILDING,"
1050 PRINT"COMPONENT LIFE, AND INTEREST  LIFE FROM"
1055 PRINT"THREE SEPARATE DISTRIBUTIONS PREVIOUSLY"
1060 PRINT"DETERMINED.  THE RESULTS  OF THE TRIALS"
1065 PRINT"ARE STORED AND ARITHMETIC MEAN,STANDARD"
1070 PRINT"DEVIATION AND  HISTOGRAM ARE CALCULATED"
1075 PRINT"FOR THE NUMBER OF TRIALS REQUESTED.":PRINT
1080 INPUT"WHAT IS INITIAL COST OF COMPONENT ";W
1085 PRINT:PRINT"HOW MANY TRIALS REQUIRED PER RUN"
1090 INPUT"               (MAXIMUM NUMBER:600) ";Q

1095 REM ************* CONDUCT REQUESTED NO OF TRIALS ************
1100 FOR N= 1 TO Q:L1=0:C=0
1105 PRINT CHR$(19);:REM                             HOMES SCREEN!!!
1110 FOR X= 1 TO 16:PRINT:NEXT X:PRINT"TRIAL NO:";N

1115 REM ***************** SELECT BUILDING LIFE ******************
1120 GOSUB 1435:R1=Y9*10
1125 IF R1>8 THEN B=(R1-2)/0.1:GOTO 1150
1130 IF R1>3 THEN B=(R1+7)/0.25:GOTO 1150
1135 IF R1>1 THEN B=(R1+1)/0.1:GOTO 1150
1140 B=R1/0.05

1145 REM ***************** SELECT COMPONENT LIFE *****************
1150 GOSUB 1435:R=Y9*10
1155 IF R>8 THEN L=(R-4)/0.4:GOTO 1175
1160 IF R>3 THEN L=(R+2)/1:GOTO 1175
1165 L=R/0.6

1170 REM ***************** SELECT INTEREST RATE ******************
1175 GOSUB 1435:R2=Y9*10
1180 IF R2>8 THEN I=(R2-4)/0.4:GOTO 1195
1185 I=(R2+5.37)/1.34

1190 REM ***************** CALCULATION OF CIU *******************
1195 L1=L1+L:IF L1>B THEN 1205
1200 C1=((W*1.1)*(1/((1+(I/100))↑L1))):C=C+C1:Z9=Z9+1:GOTO 1150
```

```
1205 T(N)=W+C
1210 NEXT N


1215 REM ************* CALCULATION OF ARITH MEAN *****************
1220 A=0:FOR J= 1 TO Q:A=A+T(J):NEXT J:A1=A/Q:A2=A1↑2


1225 REM ************* STD DEVIATION CALCULATION *****************
1230 X=0:FOR J= 1 TO Q:X=(T(J)↑2)+X:NEXT J:X1=X/Q:Z=(X1-X2)↑0.5
1235 A1=INT(.5+100*A1)/100:Z=INT(.5+100*Z)/100
1240 PRINT:PRINT"THE ARITHMETIC MEAN IS     ";A1
1245 PRINT"THE STANDARD DEVIATION IS ";Z
1250 PRINT:PRINT"DO YOU WISH TO HAVE A HISTOGRAM? (Y/N)":PRINT
1255 GET A$:IF A$<>"Y" AND A$<>"N" THEN 1255
1260 IF A$="N" THEN 1270
1265 GOSUB 1295:REM ***                   DISPLAY HISTOGRAM ROUTINE
1270 PRINT"DO YOU WISH TO TRY AGAIN? (Y/N)"
1275 GET A$:IF A$<>"Y" AND A$<>"N" THEN 1275
1280 IF A$="N" THEN DLOAD"SCREEN.GT"
1285 GOTO 1025


1290 REM ********** FIND MINIMUM & MAXIMUM OF ARRAY T() **********
1295 MI=1E7:MA=0:PRINT"PLEASE WAIT..."
1300 FOR N= 1 TO Q
1305 IF T(N)>MA THEN MA=T(N)
1310 IF T(N)<MI THEN MI=T(N)
1315 NEXT N
1320 MA=MA*1.05:MI=MI*.95:RA=(MA-MI)/10:REM*** ADD 5% & FIND RANGE
1325 FOR N= 1 TO Q
1330 K=(T(N)-MI)/(MA-MI):K=INT(10*K)+1
1335 S(K)=S(K)+1
1340 NEXT N
1345 HI=0:FOR X= 1 TO 10
1350 IF S(X)>HI THEN HI=S(X)
1355 NEXT X
1360 FQ=INT(HI/(NL-15))+1:REM ***                   FINDS FREQUENCY


1365 REM *************** HISTOGRAM SCREEN DISPLAY ****************
1370 PRINT CHR$(147);:REM                           CLEARS SCREEN!!!
1375 PRINT"HISTOGRAM CIU SIMULATION(MONTE CARLO)":PRINT
1380 PRINT"EACH * REPRESENTS A FREQUENCY OF";FQ:PRINT
1385 PRINT"--------------------------------------"
1390 P=INT(.5+100*MI)/100:FOR L= 1 TO 10:M=S(L)/FQ:N=INT(M)
1395 PRINT RIGHT$("          "+STR$(P),10);"  ";:IF N<1 THEN 1405
1400 FOR J= 1 TO N:PRINT"*";:NEXT J
1405 PRINT:P=INT(.5+100*(P+RA))/100
1410 NEXT L:PRINT"INITIAL COST      =";W;"POUNDS"
1415 PRINT"ARITHMETIC MEAN   =";A1;"POUNDS"
1420 PRINT"STANDARD DEVIATION=";Z;"POUNDS"
1425 PRINT"--------------------------------------"
1430 RETURN
1435 Y9=RND(Z9):Z9=Z9+Y9:RETURN:REM ***    GENERATES RANDOM NUMBER
```

```
1440 PRINT CHR$(147);:REM                          CLEARS SCREEN!!!
1445 PRINT"GOODBYE FOR NOW":PRINT:PRINT"END OF PROGRAM..."
1450 END
```

Chapter Eleven

Input Routines, Disk and Printer Commands

Most of the programs in this book make use of a standard set of subroutines. The purpose of this chapter is to explain their usage and how they may be altered to suit any microcomputer. Along with their explanation, equivalent statements are given for the following popular microcomputers:

BBC Microcomputer
Apple II
IBM PC
Wang PC
Intertec Superbrain
ACT Sirius
ACT Apricot
Sinclair QL*

The above list is not claimed to be comprehensive but it is hoped that it will cover most forms of BASIC available today. If the user owns a computer that is not in the list above then the machine's user manual must instead be consulted for the appropriate statement; in many cases, however, it will be similar to one of the above.

The programs themselves have been written to use the routines as frequently as possible and, as such, contain many subroutine calls. This may make the programs difficult to follow at first but once the user has read

---

* Further notes for the Sinclair QL can be found at the end of this chapter.

this chapter we hope they will become clearer. This method of programming has been adopted because it keeps the main part of the program (with two notable exceptions) as a 'universal algorithm' that may be used with any microcomputer without reducing the neatness of presentation. This has saved many man-hours in producing the disks containing all the programs in this book for the microcomputers stated on page xii, as they were transferred from computer to computer rather than laboriously typed in. The two exceptions mentioned are the coding of the commands for disk access and output to a printer. These have been left in the main program as their function cannot be easily broken down into subroutines without increasing the length of the program dramatically. Their occurrence in the programs is easily identifiable both from the listings and from the program descriptions. The statements concerned have been flagged with three exclamation marks at the end of the corresponding REMark statement and two sections are devoted to them later in this chapter.

At the end of this chapter there is a small section on testing the input routines once they have been keyed into the computer. The section includes a listing of a small program that uses every routine. It is strongly advised that this program be used to check the routines before attempting to enter any of those in the preceding chapters.

There are two final points that should be mentioned before proceeding to explain the input routines. They might seem obvious but have caused much frustration and many wasted hours. They are:

(1) Before typing anything, check to see in which case the computer accepts BASIC statements. Many computers accept both upper- and lower-case but a few accept only one of the two.

(2) Always take up-to-date copies of programs under development/being keyed in. Failure to do this is the single greatest cause of lost hours.

<u>REMEMBER:</u> If it can be lost it will be, so back it up!

There now follows a detailed description of the input subroutines:

Subroutine 4000

Function:	Clears the display screen.
Input variables:	None.
Output variables:	None.
Variables used:	CS$ Clear screen character.

Description

Line 4000 prints a character to the screen which causes it to be cleared and the cursor moved to the upper left-most position of the screen. The character to use for this in Commodore BASIC is CHR$(147). Those familiar with CBM equipment will know that this can also be represented by a 'reversed' heart or capital 'S', depending on the current display mode. This character is assigned to variable 'CS$' at the beginning of each program that uses this routine.

The cursor is kept at the top left-hand position by the semicolon at the end of the line. Otherwise the first row could never be used to display information.

Other computers use different characters, or have a BASIC statement that performs this task. Some examples are shown below:

BBC Microcomputer	CLS or PRINT CHR$(12);
Apple (Applesoft)	HOME
IBM PC	CLS
Wang PC	CLS
Superbrain	PRINT CHR$(12);
Sirius/Apricot	PRINT CHR$(27)"E";
Sinclair QL	CLS

Subroutine 4500

Function:	Homes the cursor to the upper left-most position of the screen without affecting the current display.

Input variables: None.
Output variables: None.
Variables used: HS$ Home screen character.

Description

Line 4500 prints a character to the screen which causes the cursor to go to the top left-hand corner of the screen. The character to use for this in Commodore BASIC is CHR$(19) which can also be represented as a reversed upper- or lower-case 'S'. This character is assigned to variable 'HS$' at the beginning of each of the programs that use this routine.

The semicolon ensures that the cursor remains at the home position ready to allow the next PRINT statement to display from the top left-most screen position.

Other computers have a variety of methods to obtain this function:

BBC Microcomputer	PRINT TAB(0,0);
Apple (Applesoft)	VTAB 1:HTAB 1
IBM PC	LOCATE 1,1 or PRINT CHR$(11);
Wang PC	LOCATE 1,1
Superbrain	PRINT CHR$(1);
Sirius/Apricot	PRINT CHR$(27)"H";
Sinclair QL	AT 0,0

Subroutine 5000

Function: Places the cursor at a preset position on the screen.
Input variables: RN Row number.
CL Column number.
Output variables: None.
Variables used: Z

Description

The function of this routine is to place the cursor at a specified position on the screen without altering any of the information currently being displayed. This is accomplished in the routine listed by calling the subroutine at line 4500, to home the cursor, and then moving down the screen the number of rows minus one by the use of multiple PRINTs. The final PRINT statement uses the TAB(function to tabulate the correct number of columns across the screen, the semicolon forcing the computer to remain at this position until a further PRINT statement is encountered.

This subroutine will work on most computers as far as positioning the cursor but will need to be altered on some due to a different convention being used. In Commodore BASIC, printing over a line that already has information on it using TAB(has no effect on the information being TABbed over. This is not true for many of the popular microcomputers. They overcome this problem by providing statements to display information directly to any specified position on the screen.

The alterations to the subroutine for the following microcomputers are:

BBC Microcomputer	5000 PRINT TAB(CL-1,RN-1);
Apple (Applesoft)	5000 VTAB RN:HTAB CL
IBM PC	5000 LOCATE RN,CL
Wang PC	5000 LOCATE RN,CL
Superbrain	5000 PRINT CHR$(27);"Y";CHR$(31+ RN);CHR$(31+CL);
Sirius/Apricot	As Superbrain.
Sinclair QL	5000 AT RN-1,CL-1

In each case lines 5005 and 5010 are to be deleted.

Note: You will see that for the BBC Microcomputer, 1 has to be deducted from the row and column number as the upper leftmost position is taken to be row 0, column 0, rather than row 1, column 1.

Subroutine 5500

Function:	Obtains a character from the keyboard.
Input variables:	None.
Output variables:	Z$ Contains the character typed.
Variables used:	None.

Description

This routine waits until a key is pressed and then returns the ASCII character of the key in variable 'Z$'. Unlike the INPUT statement, the character is not echoed on the screen. There are two different but similar types of BASIC statement that allow input of a character from the keyboard. One halts the program until a key is pressed, the other merely checks to see if a key has been pressed and then continues with the next statement. These statements can be made identical by adding a check to see if no key has been pressed. This check is really only necessary for the latter type but does not affect the former. The check itself compares the character input to a null or empty string and, if equal, branches back to the previous statement. (See below.) Where the BASIC concerned has both types then these have been given as alternatives.

BBC Microcomputer	5500 Z$=GET$:IF Z$="" THEN 5500 5500 Z$=INKEY$(0):IF Z$="" THEN 5500
Apple (Applesoft)	5500 GET Z$:IF Z$="" THEN 5500
IBM PC	5500 Z$=INPUT$(1):IF Z$="" THEN 5500 5500 Z$=INKEY$:IF Z$="" THEN 5500
Wang PC	As IBM PC.
Superbrain	As IBM PC.
Sirius/Apricot	As IBM PC.
Sinclair QL	5500 Z$=INKEY$:IF Z$="" THEN GOTO 5500

Subroutine 6000

Function: Inputs a number from the keyboard conforming to a specified format at a particular location on the screen.

Input variables:

- RN Row number.
- CL Column number.
- LL Maximum length of input.
- I Format type: I=0 Integer.

		I=1 Two decimal places.
	NO	Existing value.
Output variables:	NX	New value.
Variables used:	DP	Decimal point flag.
	AP	Number of characters after and including the decimal point.
	C$	String containing the value being input.
	SP$	Used to pad 'C$' with spaces.
	Z$	Temporary store for character input.
	NO	Number of characters input.
	CE$	Clear entry character.
	RM$	Abort subroutine character.
	DEL$	Backspace character.

Description

This routine allows the input of a positive number which has a program defined maximum length in characters (and therefore maximum value) and also conforms to a selected format, either integer or to two decimal places. Before this routine can be called these parameters and a few others need to be initialised to enable the routine to work correctly. The starting position of the input is specified by the variables 'RN' and 'CL'. These two variables remain constant throughout the input process and are used when the cursor position routine at line 5000 is called. This is done each time the information on the screen needs to be updated. The maximum length of the number to be input is held in variable 'LL'. This value being the number of characters the input number can take, when 'LL' is equal to four the range of values that can be input in integer mode is from 0 to 9999. This range would be reduced, if the input where in decimal mode, to 0 to 9.99 as three characters are used to specify the fractional part of the number being input. As can be seen, the type of input required must be taken into consideration when setting the value of 'LL'. The mode itself is set by the value of variable 'I' as follows:

I=0 Integer input

I=1 Decimal input to two places.

The only other variable that needs to be set is 'NO'. If the routine is to be used to allow for amendment, 'NO' is set to the existing value. If the routine is to be used for the entry of a number for the first time 'NO' is set to zero. Note that the routine does not check the format of variable 'NO' (i.e. whether it is integer or floating point) to see if it conforms to input mode specified. This is left to the programmer.

Four characters, apart from the numeric keys are accepted by the routine. These are:

(1) The 'RETURN' or 'ENTER' key. This key is used to enter the new value into the program by assigning the value of the current input to variable 'NX' and returning to the line from which the routine was called. If the routine was in decimal mode the new value is printed to two decimal places on the screen before the routine is terminated.

(2) The 'BACKSPACE' or 'DELETE' key. This key deletes one character from the right each time it is pressed, and is used to correct errors in typing. The choice of key is user-defined and is specified by initialising variable 'DEL$' to the selected ASCII value at the start of the program. The key should not be a number or letter of the alphabet but any key that is not generally used in the course of day-to-day work. For the sake of simplicity, it is suggested that the 'DEL$' variable should be assigned the value of the 'backspace' or 'delete' key on the user's keyboard. On Commodore equipment this key is character 20 (CHR$(20)) but in the case of other microcomputers it is often CHR$(8) or CHR$(127). Consult the ASCII table in the user's manual for the computer concerned.
<u>Note:</u> Apple users will have to use variable 'DE$' rather than 'DEL$' as it contains a BASIC key-word.

(3) A useful facility to have is a key which will clear the current entry completely. This can be used when the entry is no longer required or does not resemble the new

input. It is equivalent to backspacing over all the characters input until no more characters can be deleted. The choice of key is user-defined but must not be a number or letter of the alphabet. The key chosen should be assigned to variable 'CE$' at the start of each program including this routine. In the programs of the previous chapters the '@' key has been used for this purpose but the choice is left to the user.

(4) The fourth key allows the user to abort the input routine and go to another part of the main program. This key could be used to return the program to a menu or to end an input process. Again, the key to do this is user-defined but like the clear entry key should not be a number or letter of the alphabet. It is suggested that a shifted character be used (that is, one which needs the shift key to enter it). As with the previous function keys the character chosen should be assigned to variable 'RM$' at the start of the program. In the previous chapters the '↑' key has been used as it is seldom entered. Of course, this may be altered to suit the needs of the user.

In the programs of this book which use this routine, this key is used to abort the program run, clear the memory of data and start the program from the first line. This effectively re-runs the program.
Note: This facility can cause an 'out of memory' error to occur if used frequently as it does not allow the subroutine to 'return' to the calling line in the normal manner. (This problem may be resolved if the user's computer supports the 'POP' statement.)

Line Description

Lines 6000-6015	Test variable 'NO' to see if there is a previous value to be used as the input. If there is, then the number of decimal places which the number uses is found and placed in variable 'AP'. If this is more than zero then the decimal point flag 'DP' is set to one. The string

equivalent of the value is then placed in variable 'C$' padded with spaces from the variable 'SP$', and the total number of characters the value uses is calculated and replaces the value in variable 'NO'. If there is no previous value to be used, variables 'AP', 'DP', 'NO' and 'C$' are initialised to zero/spaces.

Line 6020 — Displays the current input in string 'C$' at the coordinates specified by 'RN' and 'CL'.

Line 6025 — Inputs a character from the keyboard. It is not directly echoed on the screen.

Lines 6030-6045 — Test to see if the character input is one of the four previously mentioned function keys; branching to the appropriate line if any of the tests prove true. The additional test on line 6045 ensures that there is a character that can be deleted.

Line 6050 — Tests the input character for the decimal point if in decimal mode and no decimal point has yet been input. If true then the decimal point flag is set.

Line 6055 — Tests the input character to see if it is a number; if it is not then it is discarded.

Line 6060 — Checks that the maximum number of characters has not been reached and that no more than two places of decimals are input.

Line 6065 — Ensures that the number is not greater than 9 999 999.

Line 6070 — If in decimal mode and the decimal point has not been input, the line checks to see when only three possible characters are left to be filled. At this point no more

	numeric values will be accepted until after the decimal point has been input.
Line 6075	If the decimal flag is set then variable 'AP' is incremented. This being the number of places after and including the decimal point.
Line 6080	Adds the new character to 'C$' by discarding the left-most space and increments variable 'NO' - the number of characters input. The program then branches back to line 6020.
Lines 6085-6105	Exit the routine when a carriage return has been input. The new value is found by using the VAL function on variable 'C$' and placing the result in variable 'NX'. If in decimal mode, 'C$' has a decimal point and 'zero' characters added to it where appropriate to format the display correctly.
Lines 6110-6120	Are the coding to delete a character when the backspace key is pressed. Line 6110 decrements the number of places after the decimal point (if any), and also resets the decimal point flag if it is the character to be deleted. Line 6120 discards the right-most character by adding a space to the left. The number of characters input is then decremented.

<u>Subroutine 7000</u>

Function:	Inputs a string of pre-determined maximum length.
Input variables:	RN Row number.
	CL Column number.
	LL Maximum length.
	C$ Previous string (if any).

Output variables:	C$	Contains new string.
Variables used:	NO	No. of characters input.
	SP$	Used to pad 'C$' with spaces.
	Z$	Temporary store for character input.
	CE$	Clears entry character.
	RM$	Abort subroutine character.
	DEL$	Backspace character.

Description

This routine inputs a string of characters into the program via the keyboard. The string has a maximum length defined by the value of the variable 'LL', which must be set before the routine is called. Other parameters to be initialised before calling the routine are:

(1) The co-ordinates of the starting position of the input in variables 'RN' and 'CL' for subroutine 5000.
(2) The variable 'C$'.

The latter variable should either be set to a null or empty string (C$=""), or equivalent to the string variable that requires amendment. The routine accepts the same four keys to perform special tasks as the numeric entry routine. The 'RETURN' or 'ENTER' key causes the routine to end - correctly truncating variable 'C$' - and returning to the line from which the routine was called. The other three keys work in exactly the same way as previously described.

Line Description

Line 7000	Finds the length of the string to be amended; it is then padded with spaces to the correct length.
Line 7005	Displays the string 'C$' at the position given by variables 'RN' and 'CL'.
Line 7010	Inputs a character from the keyboard. It is not directly echoed on

	the screen.
Lines 7015-7030	Test the character input to see if it is one of the four keys mentioned previously. If the 'backspace' or 'return' key has been entered then an additional test is made to ensure that there is input in string 'C$' to work with, i.e. the number of non-space characters in the string is greater than zero.
Line 7035	Discards the character input if non-printable or if the maximum length has been reached.
Line 7040	Adds the character to 'C$' by discarding the left-most space, and increments the character count. The program then branches back to line 7005.
Line 7045	Is the coding to exit from the routine after a carriage return has been input. The string 'C$' is truncated to the correct length and then returned to the main program.

Subroutine 7500

Function:	Displays a flashing error message to the screen.
Input variables:	RN Row number.
	CL Column number.
	QU No. of characters to be flashed.
	ER$ Error message.
	QU$ Prompt message.
Output variables:	None. ('CL' is altered.)
Variables used:	Z1, Z2 Used as counts.
	SP$ Pads the error message and creates 'flashing effect'.

Description

This routine flashes an error message to the screen when

an incorrect input has been received. The normal entry point is at line 7500 but any message may be displayed by presetting variable 'ER$' and then calling the routine at line 7505. The message is flashed six times (this number can be changed by altering line 7510), the duration of the flash being controlled by variable 'Z2' at lines 7520 and 7530.

Variable 'QU' contains the number of characters to be 'blinked'. This value must not exceed the column number otherwise an error will occur at line 7515. At the end of the routine the original prompting message (if any) is re-displayed ready for the appropriate input routine to be called again.

Subroutine 8000

Function:		Formats and prints a row of output to the printer/screen.
Input variables:	FO$	Format string.
	FC$	Field separator.
	N()	Numeric list to be output.
	N$()	Strings to be output.
	PR	Prints to screen if set to one.
Output variables:	FL$	Formatted string.
Variables used:	W	Field number.
	FO	Type of format required and length.
	W1	Number of preceding spaces.
	W2	Number to be formatted. (Rounded off.)
	Z$	Linefeed character, if necessary.

Description

This routine takes a list of figures, perhaps intermixed with strings and outputs them in a single string according to a defined format. The format of each field is stored in string 'FO$', its length being the number of fields to be formatted. Each character in string 'FO$' represents a certain field format and width, these parameters being found from its ASCII value. This can be explained as follows:

(Where variable 'FO' represents the selected ASCII value)

ASCII Values	Format	Field Width
FO >128	String	FO-128
FO>64	Integer (fields separated)	FO-64
FO>32	Integer	FO-32
FO>0	Decimal to two places	FO

Thus to specify a string of twenty characters followed by a number to two decimal places of ten characters the following statement is used:

```
FO$=CHR$(128+20)+CHR$(10)
```

The number of characters chosen is the maximum field width; the output values themselves are formatted by padding with spaces to this length.
The values and strings that are to be output are stored in arrays 'N()' and 'N$()' respectively before this routine is called, the element number in the appropriate array being the field number. Sample values for the above are as follows:

```
N$(1)="TOTALS":N(2)=1234.56
```

Each field is formatted and then added to the string 'FL$' until complete when the string is then output. Note that an extra character is printed after each field as a separator.

Line Description

Line 8000	Initialises the output string 'FL$' with the string 'FC$'. This string contains the character to be used as a field separator. The loop controlling the number of fields to be formatted is then set up.
Line 8010	Finds the ASCII value of a field in the format string ready for decoding.

Lines 8015-8025	Find the format type of the field and branch to the appropriate line. The default is a value to two decimal places.
Lines 8030-8035	Format a number to two places of decimals. The number is first rounded off and then the length of its integer part found; it is then added to the output string correctly padded with the appropriate number of spaces.
Lines 8045-8050	Format an integer with the field separator 'FC$'.
Lines 8055-8065	Format an integer with spaces as the field separators.
Line 8070	Formats a string (right justified).
Line 8085	If the variable 'PR' is set to one the output string 'FL$' is displayed on the screen rather than to the printer.
Line 8090	Prints the output string 'FL$' to the printer, and then returns. The variable 'Z$' can be set to a line-feed character if the printer does not line-feed automatically.

<u>Subroutine 8500</u>

Function:	Formats a positive number to two places of decimals and converts it into a string variable.	
Input variables:	B	Number to be formatted.
Output variables:	N$	Formatted string.
Variables used:	B$	Temporary store of decimal.

<u>Description</u>

This routine accepts a number in the range of 0 to 999999.99 and converts it into a string of nine characters in length and right justified, the number being formatted to two decimal places. Since three characters are needed to express the decimal part of the number, one for the decimal point and two for the decimals, this leaves six characters for the integer part of the number - hence the maximum value that can be formatted.

Line Description

Line 8500	The integer part of the number is found and right justified by converting it to a string and padding with spaces. A check is then made to see whether the number input is less than one. If true an appropriate string is set up.
Line 8505	The decimal part of the number is then found, multiplied by 100, and placed in 'B$'. A check is made for a part less than one tenth (0.1) as this requires a preceding 'zero' character to be added to the string.
Line 8510.	Adds the decimal string to the integer string, along with two extra 'zero' characters to accommodate for an integer value. The result is truncated to the first nine characters, which completes the routine.

Subroutine 9000

Function:	Pauses the program until the 'space bar' is pressed.
Input variables:	None.
Output variables:	None.
Variables used:	Z$ Input character.
	RM$ Abort subroutine character.

Description

This routine pauses the program, displaying a prompt to the user to input a 'space' character to continue the program. A check is also made to compare the character input with the character in variable 'RM$'. If they are equal then the routine is prematurely terminated, the memory cleared of data, and a jump made to the start of the program - effectively re-running the program.

Line Description

Line 9000 — Displays the prompt 'to press the space bar'. Note that two variables are printed astride the prompt. These enable the prompt to be emphasised if the computer has characters that, when printed, create special effects on the screen such as 'reverse video'. These variables must be initialised at the start of each program, either to the character(s) that perform the special effect or to 'null' or empty strings, if the computer does not support this facility. Variable 'RV$' turns the special effect on and variable 'RO$' turns it off. Computers that do not have special character sequences to create video effects may have dedicated commands to perform a similar function. A summary is listed below:

Commodore

```
RV$=CHR$(18)
RO$=CHR$(146)
```

BBC Microcomputer (Mode 7)

```
RV$=CHR$(129)+CHR$(157)+CHR$(135)
RO$=CHR$(32)+CHR$(135)+CHR$(156)
```

Apple (Applesoft)

```
RV$="":RO$=""
INVERSE            - Turns on reverse effect.
NORMAL             - Turns off reverse effect.
```

IBM PC

```
RV$="":RO$=""
COLOR 0,7            - Turns on reverse effect.
COLOR 7,0            - Turns off reverse effect.
```

Intertec Superbrain (Flashing message)

```
RV$=CHR$(27)+"~e"
RO$=CHR$(27)+"~d"
```

Each character in the message to be displayed must also have 128 added to its ASCII value. The following lines will do the conversion:

```
FOR X= 1 TO LEN(ME$)
MID$(ME$,X)=CHR$(128+ASC(RIGHT$(LEFT$(ME$,X),1)))
NEXT X
PRINT RV$;ME$;RO$
```

This will display a 'flashing' version of the contents of variable 'ME$' on the screen.

ACT Sirius/Apricot

```
RV$=CHR$(27)+"p"
RO$=CHR$(27)+"q"
```

Sinclair QL

```
RV$="":RO$=""
PAPER 7:INK 2 - Turns on reverse effect.
PAPER 2:INK 7 - Turns off special effect.
```

The above merely summarises one method of obtaining special effects on some of the more popular micro-computers. Many of these have extra commands which the user may use to 'polish' the screen display further. These have not been given in this book as it is dedicated to programs which can be freely transferred from computer to computer with the minimum of effort.

Subroutine 9500

Function:	Draws a simple 'box' around a prompt and a proposed input.	
Input variables:	RN	Row number of display.
	CL	Column number of display.
	QU$	Prompt string.
	C$	Proposed input string.
	LT$	Horizontal line string.
	LL	Maximum length of string to be input.
Output variables:	CL	Column number has length of prompt string added to it plus three to accommodate for the vertical lines.

Description

This is a simple routine to place a box around a prompt and a proposed input string to improve the screen presentation. As listed the routine uses exclamation marks for vertical lines and minus signs (contained in string 'LT$' defined at the beginning of the program) to draw the boxes. This will produce a 'rough and ready' box display on the screen but the presentation can be improved if the user's computer supports box graphics. If it is felt that this boxing effect is undesirable the routine may be ignored by replacing line 9500 with line 9540, i.e.

```
9500  CL=LEN(QU$)+3:RETURN
```

The following is a guide to the graphic character (CHR$()) codes for some of the popular microcomputers:

	CBM	BBC	IBM	ACT
Top left-hand corner	240	188	218	218
Top right-hand corner	238	236	191	191
Bottom left-hand corner	237	173	192	192
Bottom right-hand corner	253	174	217	217
Top middle vertical line	242	188	194	194
Bottom middle vertical line	241	173	193	193

Vertical lines	221	181*	179	179
Horizontal lines	192	172	196	196

* (or 234)

Notes:

1. Commodore 8032 users will have to place a line at the beginning of the program to switch into graphics mode. This is:

```
100 PRINT CHR$(142);
```

2. The BBC Microcomputer in mode 7 requires that lines which are to have graphics characters placed on them should have the character CHR$(151) placed at the beginning of each PRINT statement. This character turns on the graphic white mode. The message to be displayed should also be in capitals only.

Subroutine 9998

Function:	Ends the program run.
Input variables:	None.
Output variables:	None.
Variables Used:	None.

Description

This, as such, is not really a subroutine but merely ends the program by printing a message to this effect and then executing the END statement. Most of the programs in this book use this routine to terminate the program.

Disk Commands

The following is a description of the program alterations required for the disk access commands in Programs 2, 6B and 9, to convert them for use on a computer other than a Commodore. Also included are the commands for the loading and saving of programs which are being keyed in.

The commands have been summarised for some popular microcomputers but for others, or for a more detailed explanation of their function, refer to the computer's user guide. The statements to be replaced in the programs can be found in the Commodore BASIC 4.0 command summary.

In the text the following assumptions have been made:

(1) Where the word 'filename' occurs the user is expected to supply the appropriate name of the file to be loaded, saved or accessed sequentially. It must conform to the maximum length permitted for the user's computer, which may necessitate the changing of some of the filenames used, and also not contain the following characters:

*?",.#

(Colons should not be used unless with a drive designator.) Users of Microsoft BASIC should also avoid the use of the slash '/' and backslash '\' characters.

(2) Where (RETURN) occurs, this refers to the key to enter the command into the computer. It must not be typed in.

(3) No drive designators have been included in any of the disk commands. All commands assume the default disk drive of the computer.

(4) Where the word 'VAR' occurs, this refers to a variable to be stored or retrieved (either string or numeric).

(5) In the sequential file commands, channel number one has been used throughout as only one file is open at any time during the aforementioned programs. No attempt has been made to give details of the range of permissible channel numbers for each machine. This is left to the user if other program development is intended.

Commodore BASIC 4.0

Open read command	DOPEN#1,"filename"
Read command	INPUT#1,VAR,VAR,...
Open write command	DOPEN#1,"filename",DO,W
Write command	PRINT#1,VAR;CHR$(13);VAR;...
Close command	DCLOSE#1
Delete command	SCRATCH"filename",DO

Load command	DLOAD"filename" (RETURN)
Save command	DSAVE"filename" (RETURN)

If keying in a program, it is suggested that the following line be added, which deletes the file already on disk, re-saves it and then compares the saved version with that in memory.

```
50000 SCRATCH"filename":DSAVE"filename":VERIFY"*",8
```

It is executed by the command:

```
RUN 50000 (RETURN)
```

Filenames

CBM computers recognise filenames of up to 128 characters, but only the first 16 characters are displayed. Disk filenames must have 16 or fewer characters. Therefore it is a good idea to restrict all filenames to 16 characters or less.

<u>Commodore BASIC 2.0</u>

Open read command	OPEN1,8,5,"0:filename,S,R"
Read command	As listed.
Open write command	OPEN1,8,5,"0:filename,S,W"
Write command	As listed.
Close command	CLOSE1
Delete command	OPEN15,8,15,"S0:filename":CLOSE15
Load command	LOAD"0:filename",8 (RETURN)
Save command	SAVE"0:filename",8 (RETURN) Note: In order to save an existing file it must first be deleted.

Filenames

As for Commodore BASIC 4.0.

<u>BBC Microcomputer Model B</u>

Open read command	FILE=OPENIN("filename")
Read command	INPUT#FILE,VAR,VAR,...
Open write command	FILE=OPENOUT("filename")
Write command	PRINT#FILE,VAR,VAR,VAR,... Note: The string separator (RT$)

	shown in the listing is not needed.
Close command	CLOSE#FILE
Delete command	*DELETE"filename" This command is not neccessary when replacing files as they are over-written automatically. Furthermore, if an attempt is made to erase a file which does not exist on the diskette an error 'FILE NOT FOUND' results. This command should therefore be omitted when replacing files, but kept when solely erasing a file.
Load command	LOAD "filename" (RETURN)
Save command	SAVE "filename" (RETURN)

Filenames

Should be a maximum of seven characters and should not include any of the characters stated in note (1) on page 219.

Apple II

In the following the variable 'D$' should be initialised to: D$=CHR$(4)

Open read command	PRINT D$;"OPEN filename": PRINT D$;"READ filename"
Read command	INPUT VAR,VAR,...
Open write command	PRINT D$;"OPEN filename": PRINT D$;"WRITE filename"
Write command	PRINT VAR,VAR,...
Close command	PRINT D$;"CLOSE filename"
Delete command	PRINT D$;"DELETE filename" The same note as to usage applies here as that stated in the BBC Microcomputer section.
Load command	LOAD filename (RETURN) Note: No quotation marks needed
Save command	SAVE filename (RETURN)

Filenames

Apple II DOS allows filenames of up to 30 characters in length, but they must begin with a letter.

<u>Microsoft BASIC</u> (IBM PC, Wang PC, Sirius/Apricot, etc.)

Open read command	OPEN"I",1,"filename.TXT" Alternative form/ OPEN"filename.TXT" FOR INPUT AS #1
Read command	As listed.
Open write command	OPEN"O",1,"filename.TXT" Alternative form/ OPEN"filename.TXT" FOR OUTPUT AS #1
Write command	As listed except variable 'RT$' should be initialised to: RT$=CHR$(13)+CHR$(10) The 'PRINT#' command with the separator variable 'RT$' has been used rather than 'WRITE#' to retain compatibility with other forms of BASIC and also to allow the files to be displayed using the DOS command 'TYPE'.
Close command	CLOSE#1
Delete command	KILL"filename.filespec" Where 'filespec' is 'BAS' for a BASIC program, and 'TXT' for a sequential file. The same note as to usage applies here as that stated in the BBC Microcomputer section.
Load command	LOAD"filename" (RETURN)
Save command	SAVE"filename" (RETURN)

Filenames

Microsoft BASIC allows filenames of up to eight characters in length plus a three character file specification.

Sinclair QL

In the following, channel 3 has been used rather than channel 1 as this channel is assigned to the video display.

Open read command	OPEN_IN #3,MDV1_filename
Read command	As listed except channel 3 should be used.
Open write command	OPEN_NEW #3,MDV1_filename
Write command	As listed except channel 3 should be used. Also variable 'RT$' must be initialised to: RT$=CHR$(10)
Close command	CLOSE #3
Delete command	DELETE MDV1_filename
Load command	LOAD MDV1_filename
Save command	SAVE MDV1_filename

Filenames

Must not contain spaces but underlined spaces may be used.

Print Commands

This section explains the alterations necessary to the coding of the programs in the preceding chapters to enable the user to obtain the printouts of the results on the printer attached to his machine. Writing programs with routines to print a hardcopy of the results presents a two-fold problem. The first is that the method of opening and writing data to a channel on the printer varies from computer to computer, and the second is that there is generally a wide variety of printers that could be connected to any particular machine. The second problem has been overcome by using characters in the results which are available on all printers, such as an exclamation mark for a vertical line and a minus sign for a horizontal line. In many of the programs these characters may be changed by simply altering the assignment of the two variables allocated to them. For example, if the user has a matrix printer it may well have a vertical line in its character set which can be used instead of the exclamation mark by assigning its code to the respective variable. Also the routines do not use formatting codes to emphasise particular lines or to change the line spacing, which again would be dependent on the type of printer attached. However, an option is made available to the user to incorporate a line-feed character at the end of each printed line to allow printers which do not line-feed automatically to do so. This option can be implemented by initialising variable 'LF$' with CHR$(10), which is the line-feed character, at the beginning of the programs that have printouts. This option proves useful when the printer (which probably has a dip switch to turn the line-feed on or off) is often used with a commercial software package that requires the automatic line-feed to be off. It is left to the user to 'polish' the presentation of the printouts if so desired, but it must be stated that in doing so the routines will become less universal.

The remainder of this section is devoted to the first problem, that of converting the routines to run under a different form of BASIC. As in the previous section this is done by giving the alterations required for some popular microcomputers.

<u>Commodore BASIC 2.0</u>

Open printer channel command:

```
OPEN2,4
```

Where '2' is the channel number and '4' is the device to which the channel is opened (usually the printer).
Print data command:

```
PRINT#2,"DATA";VAR...
```

Note that the channel number must be the same as that in the open command.
Close printer channel command:

```
CLOSE2
```

The above are the same as those in the programs of the previous chapters but are included here to enable a comparison to be made with the other microcomputers.

BBC Microcomputer

Open printer channel command:

```
*FX6,0
VDU2:VDU21:Z$=""
```

These commands should be given on two separate lines to allow them to work correctly. The first command enables a line-feed character to be sent after each line. (Thus variable 'Z$' is set to a null string in the second line.) This command can be omitted if the printer is set to line-feed automatically anyway. The second line re-directs all output from the screen to the printer, the following command disabling the screen.
Print data command:

```
PRINT"DATA";VAR...
```

This is an ordinary PRINT statement. Thus, all lines with 'PRINT#2,' statements occurring in them should be changed to 'PRINT'.
Close printer channel command:

```
VDU1,10:VDU6:VDU3
```

The first command sends a line-feed character to the printer, the second re-enables the screen and the third re-directs output from the printer to the screen.
Note: The above assumes that the printer concerned uses the normal printer interface. Other interfaces may be selected using the '*FX5' command (see the BBC user guide).

Apple II

Open printer command:

```
PRINT CHR$(4);"PR#1"
```

Assuming that the printer card is in slot number one.
Print data command:

```
PRINT"DATA";VAR...
```

This is an ordinary PRINT statement. Thus all lines with 'PRINT#2,' statements occurring in them should be changed to 'PRINT'.
Close printer command:

```
PRINT CHR$(4);"PR#0"
```

Re-directs output back to the screen.

Microsoft BASIC (IBM PC, Wang PC, Sirius/Apricot etc.)

There are two possible methods of obtaining printouts when using MSBASIC. The standard method is to use the 'LPRINT' command, the other is to OPEN a file under the printer's logical device name and output to it as if it was a disk file. The latter method may only be available to users of version 2.00 or above. For the sake of completeness both are explained here, as one may suit the user's needs more precisely.

Method 1

Open printer channel command:
None needed: the statement in the listing should be replaced by a REMark, or if other statements are on the same line, omitted.
Print data command:

```
LPRINT "DATA";VAR....
```

Close printer channel command:
None needed: again the statement in the listing should be

replaced by a REMark, or if other statements are on the same line, omitted.

Method 2

Open printer channel command:
See section on disk commands. The filename must be the printer's logical device name. Typical examples are 'LPT1' for an IBM PC and 'PRN' for an ACT Sirius/Apricot.
Print data command:

```
PRINT#2,"DATA";VAR....
```

In the above example the channel number selected must be less than the maximum number of files allowed. The default number of files is usually three.
Close printer channel command:
See section on disk commands.

Sinclair QL

Open printer command:

```
BAUD 1200: OPEN#4,SER1
```

This assumes that the printer is attached to serial port 1.
Print data command:
As listed except channel 4 should be used, as channel 2 is assigned to the video display.
Close printer command:

```
CLOSE#4
```

Testing The Input Routines

In the last section of this chapter a short program is included to 'dry run' the input routines once they have been keyed in and saved. It is assumed that the user has read this chapter and has altered the routines in accordance with the suggestions made in it to suit the computer concerned. Once this short program has been keyed in and run successfully then the routines can be regarded as being correct. (When keying in the 'dry run' program there is no need to type in the REMark statements, they only serve as description.) Further errors found when the routines are used with one of the programs in the earlier chapters can therefore be traced to a probable typographical error in the main program section.

The program should be self-explanatory but a few notes are given to aid the user:

(1) For subroutine 8000, arrays 'N$()' and 'N()' have to be DIMensioned to appropriate values for the program concerned and also the function 'FNA' has to be DEFined. (See lines 1020-1025.)

(2) A standard set of variables, which are used in the input routines, need to be initialised at the beginning of every program which makes use of them. (See lines 1030-1040.)

On RUNning the program the following should occur:

(1) The screen displays 'PRESS SPACEBAR TO CONTINUE.'. This tests the routines at lines 9000 and 5500. When the user types in a space the screen should clear, testing the routine at line 4000.

(2) The user is again prompted for input and a message displayed above the previous one. This ensures the routine at line 4500 is correct.

(3) After another input the screen is cleared and two columns containing the first ten letters of the alphabet are printed down the screen. The top left-hand 'A' should then be joined to the bottom right-hand 'J' with a diagonal line consisting of the same characters to form an 'N' shape on the screen. If this is successful then the routine at line 5000 is working correctly. (Commodore users will obtain two 'A's on the left hand of the screen.)

(4) After a further input a 'flashing' message should appear briefly, thus testing the routine located at line 7500.

(5) The user is then given the opportunity to input a string (testing subroutine 7000) followed by a value

(testing subroutine 6000). This is repeated six times, allowing the user to test fully the function keys described previously in this chapter.
(6) After the sixth value has been entered the display is cleared and a table containing the six pairs of input is displayed using the subroutine at line 8000. The numeric column is then summed and the total output using the subroutine at line 8500.
(7) The test program should then terminate with a jump to the end of program routine.

If the program runs successfully the user is ready to enter those listed in the previous chapters. If, however, typing errors were made, the whole program (including the 'dry run' coding) should be re-saved until all the errors have been found and corrected. Once the user is sure that he has a working copy of the input routines then lines 1000 to 1195 may be deleted and the routines saved.

To key in a new program using these routines, first load them into the computer, then type in the main program and re-save using a different filename.

IMPORTANT!

Be sure to take back-up copies of all work as it is completed.

BASIC Differences

The functions 'LOG' and 'RND' used in Program 4 (lines 1275 and 1385) and Program 10 (line 1435) respectively may have slightly different implementations depending on the microcomputer. In Commodore BASIC 'LOG' returns the natural logarithm of a number. In other forms of BASIC this function may be called 'LN'.

The function 'RND' is available in most forms of BASIC but differences may occur in the magnitude of the number it returns depending on the value of the expression enclosed by the 'RND' function. It is left to the user to consult the computer's manual as to the correct usage of this function.

Sinclair QL Notes

The BASIC implemented on the Sinclair QL has some peculiarities which need to be considered if the reader intends to use any of the programs in this book. The following is not intended to be a complete resume of all the changes required but does give the major ones.

(1) The screen of the QL is only thirty seven characters wide by twenty rows when using a normal television as a monitor. This will cause many of the screen displays to overflow and appear untidy. The remedy is to carefully abbreviate the appropriate text and also change the values of the screen co-ordinates 'RN' and 'CL' where necessary.

(2) Concatenation (the joining of strings to each other) is done using the '&' character rather than the '+' character.

(3) The functions 'LEFT$' and 'RIGHT$' are not allowed but may be replaced as follows:

```
    LEFT$(A$,n)  changed to A$(1 TO n)
    RIGHT$(A$,n) changed to A$((LEN(A$)+1-n) TO)
```

(4) The functions 'VAL' and 'STR$' to convert a string to a number and vice versa are not required as the QL converts these when necessary without the use of these special functions.

(5) The function 'ASC' is equivalent to the function 'CODE' on the QL.

(6) The defining of functions is somewhat different from most microcomputers but here is the conversion for the function most used in this book:

```
    DEF FNA(X)=INT(0.5+100*X)/100
```

Should be changed to:

```
    DEFine FuNction FNA(X):LOCal ZZ:
    ZZ=INT(0.5+100*X)/100:RETurn ZZ:END DEFine
```

```
1000 REM ****** TEST PROGRAM FOR INPUT ROUTINES. PROGRAM 11. ******


1005 REM ********** COPYRIGHT BRANDON MOORE & MAIN. ***************
1010 REM *COLLINS PROFESSIONAL & TECHNICAL BOOKS(C)85 REV 20/04/84*
1020 DIM A$(6),A(6),N$(3),N(2)
1025 DEF FNA(X)=INT(.5+100*X)/100
1030 SP$="":LT$="":FOR X= 1 TO 79:SP$=SP$+" ":LT$=LT$+"-":NEXT X
1035 CS$=CHR$(147):HS$=CHR$(19):RM$="↑":FC$="!":RV$="<":RO$=">"
1040 DEL$=CHR$(20):CE$="@"
1050 GOSUB 9000:REM*** CHECKS ROUTINES 5500 & 9000 (PAUSE ROUTINE)
1060 GOSUB 4000:REM*** ..  ..  ..  ..  4000 (CLEAR SCREEN ROUTINE)
1065 PRINT:PRINT:PRINT"THE SCREEN SHOULD HAVE BEEN CLEARED."
1070 GOSUB 9000
1075 GOSUB 4500:REM*** CHECKS ROUTINE 4500   (HOME SCREEN ROUTINE)
1080 PRINT"THIS SHOULD APPEAR ABOVE THE PREVIOUS."
1085 GOSUB 9000:GOSUB 4000
1090 FOR X= 1 TO 10:PRINT CHR$(64+X);"          ";CHR$(64+X):NEXT X
1095 FOR X= 1 TO 10:RN=X:CL=X:GOSUB 5000
1100 PRINT CHR$(64+X):REM*** CHECKS ROUTINE 5000 (POSITION CURSOR)
1105 NEXT X:GOSUB 9000
1110 ER$="THIS MESSAGE SHOULD FLASH":RN=15:CL=37:QU=31
1115 GOSUB 7505:REM*** CHECKS ROUTINE 7500 (ERROR MESSAGE ROUTINE)
1120 FOR X= 1 TO 6:RN=18:CL=0
1125 QU$="TEXT"+STR$(X):LL=12:C$=""
1130 GOSUB 9500:REM***CHECKS ROUTINE 9500 (DRAW INPUT BOX ROUTINE)
1135 GOSUB 7000:REM***CHECKS ROUTINE 7000     (INPUT TEXT ROUTINE)
1140 A$(X)=C$:CL=25:LL=8:I=1:NO=0
1145 GOSUB 6000:REM***CHECKS ROUTINE 6000  (NUMERIC ENTRY ROUTINE)
1150 A(X)=NX:GOSUB 5000:PRINT LEFT$(SP$,10):NEXT X
1155 GOSUB 4000
1160 PRINT"!            TEXT !  VALUE    !  NO.!":PRINT LEFT$(LT$,34)
1165 FO$=CHR$(128+14)+CHR$(10)+CHR$(128+4):PR=1:B=0
1170 FOR X= 1 TO 6:N$(1)=A$(X):N(2)=A(X):N$(3)=STR$(X):B=B+A(X)
1175 GOSUB 8000:REM***CHECKS ROUTINE 8000    (FORMAT LINE ROUTINE)
1180 NEXT X
1185 GOSUB 8500:REM***CHECKS ROUTINE 8500    (FORMAT NO.  ROUTINE)
1190 PRINT:PRINT "TOTAL=";RIGHT$(SP$+N$,20)
1195 GOTO 9998:REM*** CHECKS ROUTINE 9998    (END PROGRAM ROUTINE)
```

```
3997 REM *********************** SUBROUTINES **********************

3998 REM *********************************************************

3999 REM ***************** CLEAR SCREEN COMMAND *******************
4000 PRINT CS$;:REM                                 CLEARS SCREEN!!!
4005 RETURN

4499 REM ***************** HOME SCREEN COMMAND ********************
4500 PRINT HS$;:REM                                  HOMES SCREEN!!!
4505 RETURN

4999 REM *************** CURSOR POSITION ROUTINE ******************
5000 GOSUB 4500:IF RN=1 THEN 5010:REM                HOMES SCREEN
5005 FOR Z= 1 TO RN-1:PRINT:NEXT Z
5010 PRINT TAB(CL);
5015 RETURN

5499 REM ************ INPUT A CHARACTER FROM KEYBOARD *************
5500 GET Z$:IF Z$="" THEN 5500
5505 RETURN

5999 REM ************* INPUT A NUMERIC VALUE ROUTINE **************
6000 IF NO=0 THEN 6015
6005 DP=0:AP=LEN(STR$(NO))-LEN(STR$(INT(NO))):IF AP>0 THEN DP=1
6010 C$=RIGHT$(SP$+STR$(NO),LL):NO=LEN(STR$(NO))-1:GOTO 6020
6015 AP=0:DP=0:NO=0:C$=RIGHT$(SP$,LL)
6020 GOSUB 5000:PRINT C$;"<"
6025 GOSUB 5500:REM                                 INPUT CHARACTER
6030 IF Z$=CE$ THEN 6015
6035 IF Z$=RM$ THEN CLR:GOTO 1000:REM               CLEARS MEMORY!!!
6040 IF Z$=CHR$(13) THEN 6085
6045 IF Z$=DEL$ AND NO>0 THEN 6110
6050 IF Z$="." AND I=1 AND NO<LL AND DP=0 THEN DP=1:GOTO 6060
6055 IF VAL(Z$)=0 AND Z$<>"0" THEN 6025
6060 IF NO=LL OR AP>2 THEN 6025
6065 IF NO=7 AND DP=0 THEN 6025
6070 IF I=1 AND DP=0 AND NO>=LL-3 THEN 6025
6075 IF DP=1 THEN AP=AP+1
6080 C$=RIGHT$(C$,LL-1)+Z$:NO=NO+1:GOTO 6020
6085 IF I=0 THEN 6100
6090 IF DP=0 THEN C$=C$+".":DP=1:AP=1
6095 C$=C$+LEFT$("00",3-AP):C$=RIGHT$(C$,LL)
6100 NX=VAL(C$):GOSUB 5000:PRINT C$;" "
6105 RETURN
6110 IF DP=1 THEN AP=AP-1:IF AP=0 THEN DP=0
6120 C$=" "+LEFT$(C$,LL-1):NO=NO-1:GOTO 6020
```

```
6999 REM *************** INPUT ALPHANUMERIC ROUTINE ****************
7000 NO=LEN(C$):C$=RIGHT$(SP$+C$,LL):IF C$=LEFT$(SP$,LL) THEN NO=0
7005 GOSUB 5000:PRINT C$;"<"
7010 GOSUB 5500:REM                                 INPUT CHARACTER
7015 IF Z$=CE$ THEN C$="":GOTO 7000
7020 IF Z$=RM$ THEN CLR:GOTO 1000:REM               CLEARS MEMORY!!!
7025 IF Z$=CHR$(13) AND NO>0 THEN 7045
7030 IF Z$=DEL$ AND NO>0 THEN C$=" "+LEFT$(C$,LL-1):NO=NO-1
7035 IF NO=LL OR ASC(Z$)<32 OR ASC(Z$)>96 OR Z$=DEL$ THEN 7005
7040 C$=RIGHT$(C$,LL-1)+Z$:NO=NO+1:GOTO 7005
7045 GOSUB 5000:PRINT C$;" ":C$=RIGHT$(C$,NO)
7050 RETURN


7499 REM ***************** REPORT ERROR ROUTINE ******************
7500 ER$="ILLEGAL ENTRY"
7505 CL=CL-QU
7510 FOR Z1= 1 TO 6
7515 GOSUB 5000:PRINT LEFT$(ER$+SP$,QU)
7520 FOR Z2= 1 TO 100:NEXT Z2
7525 GOSUB 5000:PRINT LEFT$(SP$,QU)
7530 FOR Z2= 1 TO 100:NEXT Z2
7535 NEXT Z1
7540 GOSUB 5000:PRINT LEFT$(QU$+SP$,QU):RETURN


7999 REM ***************  FORMAT AND PRINT LINE  *****************
8000 FL$=FC$:FOR W= 1 TO LEN(FO$)
8010 FO=ASC(LEFT$(RIGHT$(FO$,LEN(FO$)-W+1),1))
8015 IF FO AND 128 THEN 8070:REM *** FORMAT STRING
8020 IF FO AND 64 THEN 8045:REM  *** FORMAT INTEGER
8025 IF FO AND 32 THEN 8055:REM  *** FORMAT INTEGER WITH SPACES
8030 W2=FNA(N(W)):W1=FO-4-LEN(STR$(INT(W2)))
8035 FL$=FL$+LEFT$(LEFT$(SP$,W1)+STR$(W2)+SP$,FO)+FC$:GOTO 8080
8045 FO=FO-64:W2=INT(.5+N(W)):W1=FO-1-LEN(STR$(INT(W2)))
8050 FL$=FL$+LEFT$(LEFT$(SP$,W1)+STR$(W2)+SP$,FO)+"!":GOTO 8080
8055 IF W=1 THEN FL$=FC$
8060 FO=FO-32:W2=INT(.5+N(W)):W1=FO-1-LEN(STR$(INT(W2)))
8065 FL$=FL$+LEFT$(LEFT$(SP$,W1)+STR$(W2)+SP$,FO)+" ":GOTO 8080
8070 FL$=FL$+RIGHT$(SP$+N$(W),FO-128)+" "+FC$
8080 NEXT W
8085 IF PR=1 THEN PRINT FL$:RETURN
8090 PRINT#2,FL$;Z$:RETURN


8499 REM ************* FORMAT A NUMBER TO 2 PLACES DEC.************
8500 N$=RIGHT$("      "+STR$(INT(B)),6)+".":IF B<1 THEN N$="     0."
8505 B$=STR$(INT(.5+100*(B-INT(B)))):IF VAL(B$)<10 THEN N$=N$+"0"
8510 N$=N$+RIGHT$(B$,LEN(B$)-1)+"00":N$=LEFT$(N$,9):RETURN


8999 REM ************ PRESS A KEY TO CONTINUE ROUTINE *************
9000 PRINT:PRINT"    ";RV$;"PRESS 'SPACE BAR' TO CONTINUE";RO$;
```

```
9010 GOSUB 5500:IF Z$<>" " AND Z$<>RM$ THEN 9010
9020 IF Z$=RM$ THEN CLR:GOTO 1000:REM            CLEARS MEMORY!!!
9030 RETURN


9499 REM ******************* DRAW INPUT BOXES *********************
9500 RN=RN-1:GOSUB 5000:RN=RN+1
9510 PRINT"!";LEFT$(LT$,LEN(QU$)+1);"!";LEFT$(LT$,LL+1);"!"
9520 PRINT"!";QU$;" !";RIGHT$(SP$+C$,LL);" !"
9530 PRINT"!";LEFT$(LT$,LEN(QU$)+1);"!";LEFT$(LT$,LL+1);"!"
9540 CL=LEN(QU$)+3:RETURN


9997 REM ***************** END PROGRAM ROUTINE ********************
9998 PRINT:PRINT"END OF PROGRAM...."
9999 END
```